Both ENDS is an Amsterdam-based service and advocacy organisation which collaborates with hundreds of environmental and indigenous organisations, both in the South and in the North, with the aim of helping to create and sustain a vigilant and effective environmental movement.

Econet is a voluntary organisation based in Pune, India, working in the field of ecological networking, environmental restoration and social justice. Established in 1989, it has worked for the protection of community and tribal lands through public interest litigation, and also by supporting forest management and reforestation programmes. It has prepared river-basin and micro-watershed plans, and lobbied for the restoration of traditional water-harvesting systems in western India through participatory action research.

Zed Titles on Forestry

As part of our wide-ranging Development and the Environment list, we have published extensively on forestry issues – covering not only the threats to and destruction of the world's forests, but also the traditional and innovative practices oriented towards their sustainable management and use. Titles include:

Kojo Amanor
The New Frontier: Farmers' Responses to Land Degradation

Tariq Banuri and Frederique Atffel-Marglin (eds)
Who Will Save the Forests?
Knowledge, Power and Environmental Destruction

Riccardo Carrere and Larry Lohmann
Pulping the South
Industrial Tree Plantations and the World Paper Economy

Marcus Colchester and Larry Lohmann (eds)
The Struggle for Land and the Fate of the Forests

Bertus Haverkort and Wim Hiemstra
Food for Thought
Ancient Visions and New Experiments of Rural People

Kgathi, Hall, Hategeka and Sekhwela,
Biomass Energy Policy in Africa

John Overton and Regina Scheyvens (eds)
Strategies for Sustainable Development: Experiences from the Pacific

Peter Read
Responding to Global Warming: The Technology, Economics and Politics of Sustainable Energy

Peter Stone (ed.)
The State of the World's Mountains: A Global Report

Bill Weinberg
War on the Land: Ecology and Politics in Central America

Paul Wolvekamp (ed.)
in collaboration with
Ann Danaiya Usher, Vijay Paranjpye and Madhu Ramnath:
Forests for the Future: Local Strategies for Forest Protection, Economic Welfare and Social Justice

For full details about these titles and Zed's general and subject catalogues, please write to:

The Marketing Department, Zed Books,
7 Cynthia Street, London N1 9JF, UK or
email Sales@zedbooks.demon.co.uk

Visit our website at: http://www.zedbooks.demon.co.uk

Forests for the Future

Local Strategies for Forest Protection, Economic Welfare and Social Justice

edited by

Paul Wolvekamp

in collaboration with

Ann Danaiya Usher

Vijay Paranjpye

Madhu Ramnath

Zed Books

LONDON & NEW YORK

Forests for the Future: Local Strategies for Forest Protection, Economic Welfare and Social Justice
was first published in 1999 by
Zed Books Ltd, 7 Cynthia Street, London, N1 9JF and
Room 400, 175 Fifth Avenue, New York, NY 10010, USA

in association with
Both ENDS, Damrak 28–30, 1012 LJ, Amsterdam, The Netherlands
and
Econet, 4 Pushpali Apartments, Wrangler Paranjpye Road, 1218/3 Shivaji Nagar,
Pune – 411 004, India

Cover design by Andrew Corbett
Set in 10/11 pt Berkeley Book
by Long House, Cumbria, UK
Printed and bound in the United Kingdom
by Biddles Ltd, Guildford and King's Lynn

A catalogue record for this book
is available from the British Library

ISBN   Hb 1 85649 756 9
       Pb 1 85649 757 7

Publication of this book has been made possible through financial support from the Tropical
Rainforest Programme of the Netherlands Committee of the International Union for the
Conservation of Nature (NC–IUCN). The NC–IUCN manages a Dutch Government small
grants programme which finances small-scale NGO projects promoting the conservation
and effective management of tropical rainforests. NC–IUCN is a membership organisation
whose members are also members of the IUCN, which promotes similar aims in an inter-
national context.
Address: Plantage Middenlaan 2–B, 1018 DD Amsterdam.
Telephone 31-20-6261732; Fax 31-20-6279349.
Email: mail@nciucn.nl
Website: www.nciucn.nl

•••

The Directorate General XI of the European Commission, the Tropical Rainforest
Programme of the Netherlands Committee for IUCN, the Forest Emergency Fund, the
Stichting School Koning Willem III and Koningin Emma Fonds provided financial support
to conduct the case studies and to write this book.

Contents

Note on the Contributors

The Editors

Paul Wolvekamp has an educational background in law, political science and environmental studies. His work required him to travel extensively in India and other southern countries. He has been an affiliate of the Amsterdam-based foundation Both ENDS for nearly ten years. His main professional interest is in ways to achieve recognition for the realities and priorities of local people and their support organisations with regard to environmental management and 'development' *vis-à-vis* the approaches of development cooperation and the market economy.

Ann Danaiya Usher is a Thai-Canadian journalist who has been a staff writer at the *Nation* newspaper in Bangkok, and now works for the Oslo-based journal *Development Today*. She is the author of *Dams as Aid: a Political Anatomy of Nordic Development Thinking* (Routledge 1997), and is currently living in Norway and writing a book on Thai forest politics.

Vijay Paranjpye has been teaching economics at the Ness Wadia College, University of Pune, and is the founding director of the NGO Econet and a trustee of the organisations Gangotri and Gram Vardhini. A pioneer in the field of scientific cost-benefit analysis using environmental criteria, he has analysed and written books on several controversial projects like the Tehri Dam and the high dams on the Narmada River. He has advised the government of Maharasthra and Madhya Pradesh on ecologically and socially sensitive projects and has testified before the US Congress. Vijay Paranjpye continues to write on a range of environmental subjects.

Madhu Ramnath has a background in mathematics and philosophy and is trained as a botanist. For more than fifteen years he has worked with and lived among *adivasi* (tribal) communities in most states in India. His work concentrates on ethnobotanical action research with the further aim of gaining wider recognition for the values and knowledge of, and the problems faced by, forest-dwelling *adivasi* communities.

The Authors of the Case Studies

Accion Ecologica is a campaign organisation working with local communities on issues that include forest conservation, mining, oil exploration and indigenous land rights.

Arbofilia is a farmers' cooperative-cum-NGO which works towards ecological and cultural restoration in Costa Rica, with emphasis on efforts which yield tangible results in landscape recovery, biodiversity conservation, water protection and social services.

Jenne de Beer conducted one of the first comprehensive field-level surveys on the role of non-timber forest products (NTFPs) in Southeast Asia and the Russian Far East, notably their immediate importance for local families and for forest conservation. He is field coordinator of the NTFP Exchange Programme for Southeast Asia and collaborates closely with local communities and NGOs in the fields of land and user rights and sustainable exploitation and marketing of NTFPs. He is an anthropologist and co-author of the book *The Economic Value of Non-Timber Forest Products in Southeast Asia* (NC-IUCN, 1997).

Centre d'Action pour le Développement Durable et Intégré Dans les Communautés (CADIC), an NGO in the Democratic Republic of Congo, is a pioneer in the field of people-oriented nature conservation in Central Africa, notably on the eastern border of the Democratic Republic of Congo. For many years CADIC staff have been working with local village communities and nomadic Pygmy groups.

Centre pour l'Environnement et le Développement (CED) works closely with Baka (Pygmy) and Bantu communities in densely forested south Cameroon and endeavours to help protect and restore their rights over forest through the promotion of community forestry, legal training, advocacy work and campaigns against destructive interventions such as illegal logging, plantations, road building and other infrastructural projects.

Conservation Council of New Brunswick is a citizen-based environmental organisation that has been campaigning for a transition from centrally planned forest use to community forestry in this province on the east coast of Canada.

Russel Diabo is a Mohawk Indian who worked with the Algonquins of Barrier Lake, Ontario as an adviser on land rights and forest management issues.

Environmental Defense Fund is a large US-based environmental advocacy organisation working on a range of domestic and international issues, including forest management and local-indigenous rights.

The Confederated Tribes of Warm Springs in central Oregon, are made up of the Wasco, Warm Springs and Paiute tribes and number about 4,000 members. Historically the tribes have lived throughout the Columbia River basin, depending on the salmon fisheries.

ETC Netherlands, a non-profit consultancy firm, aims to encourage and support local initiatives in sustainable development, notably in the fields of sustainable agriculture, agro-forestry, energy, water supply and institutional development.

Forest Group Limburg in southern Netherlands is a cooperative association formed by forest owners, municipalities and nature conservation organisations at the provincial level with the aim of promoting improved and more efficient management of its members' forests.

Green Earth Organisation in Ghana was established by a group of young professionals to halt the outbreak of forest fires near the national capital. The organisation gradually assumed the broader task of promoting the improvement of local environmental management through training, dialogue and advocacy.

Montanosa Research and Development Centre, the Philippines, is a service and advocacy organisation in support of indigenous communities in the Cordillera on Luzon Island.

Nature's Beckon in Assam, India, founded more than 15 years ago by a group of young nature lovers, evolved into a locally rooted organisation which together with local tribal communities assumed responsibility for the protection and management of an endangered national park.

Polish Ecological Club is a national advocacy organisation involved in research, campaigns and consultations on such issues as forest management, energy, transport, agriculture and economic development.

Prakruthi is a social activist and campaign organisation which stems from the Appiko/Chipko grassroots mass movement in South India's Western Ghats. It focuses on local peoples' rights over forest and the abatement of ecological destruction by mining, logging, nuclear plants, energy and transport infrastructure projects.

Boyce Richardson is an Ottawa writer and film maker who has worked with aboriginal people in Canada for 25 years. One of his most recent books is *People of Terra Nullius*, an historical and contemporary account of Canadian Indian nations.

Sobrevivencia (Friends of the Earth Paraguay) works on a range of local, national and international social and environmental issues and collaborates closely with forest-dependent indigenous communities and farmers in Paraguay. It runs a model farm and assists local communities in their struggle for land rights, immediate survival and the general improvement of their standard of living.

Vitae Civilis (Brazil) consist of a team of professionals with a background in areas such as law, medicines, engineering, forestry and botany. It works closely with communities in Mata Atlantica forest in Sao Paulo State, as well as with policy makers, NGOs and other parties at the state and national level towards legal reform in favour of forest protection and the recognition of local peoples' rights over and knowledge of forest and biodiversity.

Andrey Zakharenkov is director of the Far Eastern Association for the Use of Non-Timber Forest Products in Russia. He has an extensive experience in promoting sustainable NTFP production and community support. The Association, based in the town of Khabarovsk, has as its primary task to ensure sustainable development in remote forest villages and to support the traditional economies of the region's indigenous peoples in NTFP trade.

Abbreviations

ADB	Asian Development Bank
AECO	Asociacion Ecologista Costarricense
ASUP	Asosasyon dagiti Sosyudad ti Umili ti Pidlisan (Pidlisan people's association, Philippines)
BIA	Bureau of Indian Affairs
BLM	Bureau of Land Management
BMP	Best management practice
CADIC	Centre d'Action pour le Développement Durable et Intégré Dans les Communautés
CBS	Centraal Bureau voor de Statistiek
CCNB	Conservation Council of New Brunswick
CED	Centre pour l'Environnement et le Développement
CIFOR	Center for International Forestry Research
CIRAD	Centre de Cooperation International en Recherches Agronomiques pour le Développement
CO_2	Carbon dioxide
CONAIE	Confederation of Indigenous Nationalities of Ecuador
DGIS	Department for International Cooperation, Netherlands
ECOFAC	Le Programme de Conservation et d'Utilisation Rational des Ecosystèmes d'Afrique Centrale
EDF	Environmental Defense Fund
ELCI	Environment Liaison Centre International
EPA	Environmentally protected area
ETC	Educational Training Consultants
EU	European Union
FECCHE	Federation of Chachi Centres of Esmeraldas
FEPP	Fondo Ecuatoriano Populorum Progressio
FOE	Friends of the Earth
FUNDEAL	Fundacion para el Desarrollo Alternativo

GATT	General Agreement on Tariffs and Trade
GEF	Global Environmental Facility
GEO	Green Earth Organisation
GIS	Geographical information systems
IBR	Rural Welfare Institute (Paraguay)
ID	Interdisciplinary (Team)
IKL	(Stichting) Instandhouding Kleinschalige Landschapselementen (Foundation for the Preservation of Small-Scale Landscapes
ILO	International Labour Organisation
IMF	International Monetary Fund
INEFAN	Instituto Ecuatoriano Forestal de Areas Naturales y Vida Silvestre
IRMP	Integrated Resources Management Plan
IUCN	International Union for the Conservation of Nature
MAI	Multilateral Agreement on Investment
NATRIPAL	United Tribes of Palawan
NSRC	Neo Synthesis Research Centre
NGO	Non-governmental organisation
NTFP	Non-timber forest products
OECD	Organisation for Economic Cooperation and Development
PLANFOR	Plan Forestal
PROFAFOR	Programa Face de Forestacion
RH	Rimbunan Hijau
SAP	Structural adjustment programme
SINASIP	National System of Protected Wild Areas
SLAPP	Strategic Lawsuits Against Public Participation
TRIPs	Trade-Related Aspects of Intellectual Property Rights
US	United States
WHO	World Health Organisation
WWF	World Wildlife Fund
ZSI	Zoological Survey of India

Acknowledgements

In 1992, Both ENDS and Econet started facilitating the recording of a range of case studies on local-indigenous forest management in tropical, temperate and boreal countries by local organisations and individuals. This book compiles the insights and experiences generated by these studies.

During the preparation of the case studies and the compilation and editing of this book valuable advice and support was received from a great number of people. We want to express profound appreciation to all.

In particular we wish to thank the authors of the case studies, with their colleagues and constituencies, whose trust and dedication made the book possible. We express our gratitude to: Joshua Awaku-Apaw, Jenne de Beer, Matthew Betts, Gemima Cabral Born, Rubens Born, Nicole Cordewener, David Coon, Irma Corten, Soumyadeep Datta, Florence Daguitan, Russel Diabo, Elias Peña Diaz, Swedi Elongo, Lorena Gamboa, Pandurang Hegde, Bogumila Kuklik, Annabelle Maffioli Reyes, Deborah Moore, Samuel Nguiffo, Lambert Okrah, Boyce Richardson, Jerzy Sawicki, Deepak Sehgal, Rafal Serafin, S. T. Somasekhare Reddy, Katarzyna Terlecka, Thomasz Terlecki and Andrey Zakharenkov.

A word of thanks, too, to Patrick Anderson, Jean Arnold, Bill Barclay, Kees van Dijk, Ton van Eck, Chris Genovani, Peter Laban, Maria Athena Ronquillo and colleagues at Both ENDS, notably Tamara Mohr, Bert Zijlstra and Marie-Jose Vervest. They introduced us to individuals and organisations who enriched our own understanding of the realities on the ground and who brought a tangible contribution to this book.

Marcus Colchester, Martin von Hildebrand, Liz Hosken, Calestous Juma, Srisuwan Kuanchachorn, Karen Lindahl, Wangari Mathaai, Wittoon Per-Pongsacharoon, Ranil Senanayake and Miguel Soto Cruz gave valuable advice on the project, while support during it came from Babette Graeber, Anne Janssen, Theo van Koolwijk, Suman Natarajan, Roger Olsen, Jeanette van Rijsoort, Jan Stevens, Stefan Vandenbijlaard and Inger van de Werf.

The Directorate General XI of the European Commission, the Tropical Rainforest Programme of the Netherlands Committee for IUCN, the Forest Emergency Fund, The Stichting School Koning Willem III and Koningin Emma Fonds provided financial support for conducting the case studies and the writing and publishing of this book. It is worth mentioning, however, that the spin-off of these grants is much greater, since it helped many NGOs and communities to obtain wider recognition and support for their cause. In particular we wish to thank the following people for their enthusiasm and patience: Wim Bergmans, Annelies van Dam, Wouter van Dam, Willem Ferwerda, Rietje Grit, Pierre Hamoir, Ron Kingham, Jaap Kuper, Bernard Mallet, Julio Ruiz Murieta, Machiel Overhof, Sian Petman and Wouter Veening.

We would like to thank all those who offered valuable comments and assistance during the writing and final preparation of the manuscript. Jasper Altena deserves great appreciation for his relentless support and meticulous work. Gerhard van den Top, Larry Lohmann, Bob Ursem and Anjali Paranjpye made very constructive comments and textual suggestions. We thank Katherine and Mike Kirkwood for book production. Thanks are also due to Jenne de Beer, Cas Besseling, Ann Doherty, Willy Douma, Rod Harbinson, Erica Koning, Bert-Jan Ottens, Eline Meyer, Wiert Wiertsema and Laurian Zwart.

Finally, a special word of appreciation is reserved for Robert Molteno of Zed Books, who has shown an explicit interest in the subject right from the beginning.

Preface

It is estimated that the world's forests are vital for the daily survival of more than 300 million indigenous and peasant people who depend on forest ecosystems (World Bank Forest Policy 1990). These communities have devised sophisticated norms for managing watersheds, catchment areas and fragile forest ecosystems, and possess a wealth of knowledge about rational land use and environmental protection. Many such rural communities are important forest stakeholders. Yet the expertise and interests of these local people are rarely recognised by national forest policies and management systems. They are often accused of being the main agents of forest destruction, and their position is further marginalised. Instead, government institutions tend to be viewed as the principal actors in forest conservation and restoration.

In many countries central government claims control over forest resources, largely ignoring the customary rights of forest communities and thus eroding traditions, responsibilities and decision-making structures at the local level. Western 'scientific' forestry, introduced world-wide in the course of the twentieth century, has been very influential in this respect. This brand of forestry usually neglects and often undermines local forestry systems. Forced resettlement, for example, is perceived as a prerequisite for watershed and park protection. Concessions for commercial logging are provided without proper consultation. Tree plantations that fulfil national reforestation goals replace farmland and sometimes even natural forest, threatening local biological diversity.

Much of the international discussion of forests – tropical forests, in particular – has focused on the biological diversity crisis. Yet the spectre of massive global deforestation also represents a grave threat to human communities. Many forest-dependent communities – whether forest-dwelling ethnic minorities or farmers who rely on a patch of secondary forest for subsistence – lack both land security and political representation. These people are, so to speak, at the front line. They face pressure from outsiders who seek land, timber or other resources; they are exposed to intimidation, violence and culture shock; and they confront internal problems about balancing forest exploitation and conservation. They are often torn between maintaining a forest area as a watershed for their fields and market pressure to cut timber for profit. Consequently, forest communities are blamed for deforestation and ecological degradation of forest areas, and are regularly accused of being incapable of managing their own forest lands.

Local non-governmental and grassroots organisations can rarely devote time and resources to analysing and documenting their experiences and point of view for larger audiences. Existing studies on community forest management seldom lead to policy conclusions, or benefit local stake-holders and their causes. Conscious of these realities, Both ENDS and Gram Vardhini embarked on a collaborative survey project in 1992. The objective was to enable forest communities to bring to public attention their own perceptions and experiences. They would describe in their own words how they are striving to balance cultural and economic survival with sustenance of the ecosystems on which they depend, under pressure from a growing population, increasing demands for cash, and a range of outside forces.

The initiators of the survey had been concerned about the tendency to locate the problems of deforestation and biological diversity depletion exclusively in Southern countries, even though forest-dependent communities in the industrialised world are also at risk. It was therefore important that organisations from the temperate and boreal regions joined the survey. This book is thus a collection of case studies undertaken in many corners of the world, under a variety of ecological and socio-economic circumstances.

The case studies show how community control and involvement can allow for more detailed assessments of forest resources and management needs than centralised forest management. Local communities often have a very long history of using forest produce and regulating access to forest resources. There still exists at the local level an enormous variety of struc-tured ownership arrangements, incentives and sanctions that work to ensure compliance. Given the impasse in international forest negotiations and the inertia of most governments, it is important to consider the alternatives. Better understanding will provide greater support for local citizens' initia-tives to sustain forest resources.

Through the compilation of documented evidence, the studies reveal that at times local forest management has benefited from moral, technical, political and financial support from outsiders – NGOs, scientists, consul-tants, journalists and government or donor agencies. Often, however, local groups work in isolation. The case studies describe concrete situations that embody what the authors and their constituencies observe, believe and strive for. The essays challenge the notion that forest communities are problems, while state bodies deliver solutions. Unavoidably, some texts are unpolished. Using their political and social instincts, the authors go to the heart of the matter, avoiding scientific or wordy speculations.

These testimonies may help to underline the need for outsiders to be more sensitive to local interests and perspectives. We will be encouraged if this collection of essays motivates other local organisations to put their ideas on paper in the cause of sustained local forest management. And we hope it will help national governments and international donor agencies to appreciate local peoples' capacities and views on forest management, stimulating greater collaboration with local organisations and their support groups.

Paul Wolvekamp

Prologue

When the ruler's trust is wanting,
there will be no trust in him.
Cautious,
he values his words.
When his work is completed and his affairs finished,
the common people say,
We are like this by ourselves.

Lao Tzu[1]

The objective of this book is to enable local people to document and present their own views and experiences of local forest management to a wider world. The book is a result of a joint project, a survey of forest management by indigenous people and other local populations in tropical, temperate and boreal[2] countries. It was preceded by a long process of collaboration between a great many individuals and organisations. The sum of evidence from these different case studies should generate more recognition of local forest management systems and their potential to sustain local economies and to preserve much of the world's remaining forests. Moreover, the local organisations that participated believe that their own work on the ground will benefit from such action research.

All the case studies therefore address the same key question: *how can local/indigenous communities maintain the balance between their societies and forest environments when faced with rising populations, growing demands for basic needs and cash, and increasingly stronger external pressures?*

Virtually all the case studies witness deforestation, economic blunders and social injustice. Local forest management practices in most parts of the world are clearly under increasing physical and psychological pressure. Despite very different ecological, political and economic circumstances, it is easy to establish common causes of forest destruction and the loss of local livelihoods and culture. Unequal access to forest resources is the most important of these. Forest areas world-wide host major reservoirs of minerals,

1

metals, biomass, land for agricultural expansion and other resources. Most case studies report conflicts over these resources since their national political and economic elites are unwilling to forgo the opportunity to tap these reservoirs, notwithstanding the often dramatic social and environmental consequences. The case studies also confirm that lack of security of land rights and user rights is a major cause of decline in local systems of forest management, resulting in social hardship and forest destruction. It is also clear that few democratically elected governments in Northern or Southern countries are enthusiastic about sharing control and rights over forests with local communities. In many instances, there is overt collusion between government agencies and dominant economic interest groups. One observes, for example, the granting of extensive privileges – such as mining or logging concessions, subsidies and tax exemptions – to a small number of industrial conglomerates.

Yet the real life experiences compiled in this book also help us to identify a number of unique responses, perceptions and practices by local people and other concerned parties. And it is also possible to translate some of these special insights into more general conclusions and policy recommendations. This book has the purpose of communicating these findings to those parties whose policies and actions have a direct impact on local forest management: decision makers, donor agencies, corporations, researchers, non-governmental organisations (NGOs), the media and the public at large. It indicates how they could open more space for the enhancement of prudent and undisturbed management of forests by local people.

This chapter is organised in three sections. The first section deals with the potential of local forest management systems, probing their social and institutional strengths and weaknesses. It also responds to prevailing scepticism about these systems. The second section identifies the main causes of their collapse. The third section presents the key lessons to be learned from the case studies. It draws some general conclusions and has specific recommendations for policy makers, donors, researchers and other groups of players.

Local Forest Management under Scrutiny

Predictably and invariably, forest industries and other commercial interests have opposed the legitimisation of forest management by local communities. But government authorities, media and academic institutions have also questioned the ability of local people to manage their resources prudently. Critics express their scepticism by pointing out that:

1   Local forest users are not capable of coping with changing socio-demographic and economic circumstances, or with the new demands on forest management;

2 Local forest management does not safeguard conservation interests adequately;

3 Local forest users are unable to resist external sources of degradation and fail to restore degraded forest land;

4 Local communities feature social and economic inequalities and institutional weaknesses which frustrate sustainable forest management (Carrere and Lohmann 1996; Colchester 1992; .Jepma 1995).

It is impossible to generalise about the commitment and capacity of local people to preserve forest and biodiversity. Among the hundreds of millions of villagers who live in close connection with their local forests, it is often the indigenous peoples who maintain a relatively non-agricultural, non-market relationship with the forest. Hunters and gatherers such as the Durva in central India – whose custom it is to pass their lands on, unharmed, to the generations that follow them – manage their resources cautiously in order to ensure a sustained yield. Shifting cultivation[3] practices by indigenous communities, for instance, reveal not only the extreme variability and complexity of these traditional technologies, but also the enormous reserve of vernacular knowledge of practices to restore soil fertility and to preserve biodiversity (Perpongsacharoen and Lohmann 1989; Colchester 1992; see also Colfer and Dudley 1993). The Durva songs about pollination illustrate this point well.[4]

Not all forest-dependent people are members of ethnic minorities. In the most frequent case a patch of secondary forest is part of the subsistence guarantee for the poorer section of the village. The forest provides fodder, cropland, protein, medicine, firewood, mushrooms, vegetables, building materials or any number of other products. Not unlike the indigenous groups, many peasant peoples – even those whose main economic activity is permanent agriculture – have a very long history of using forest produce and regulating access to forest resources. 'There exists an enormous variety of structured ownership arrangements within which management rules are developed, group size is known and enforced, incentives are in place for co-owners to follow the accepted institutional arrangements, and sanctions work to ensure compliance' (Cernea 1989 in Colchester 1992: 120). It must be acknowledged, however, that local management in varying degrees manipulates the forest to satisfy local needs and hence it affects the pristine state of the forest's ecology (Hildyard *et al.* 1997).

The case studies give evidence that environmental decline in forest areas inhabited by or adjacent to local communities often occurs where local social institutions and the environment are simultaneously under heavy pressure from the outside. Many of these people share a lack of both land security and political influence. They live, so to speak, at the 'front line'. Poverty and population growth – and the corresponding local demand for food and other basic needs – certainly increased pressure on the local forest environment in many regions. And yet, in many areas, 'overpopulation' is a misleading concept if one takes land distribution into account. Areas which are deemed 'overpopulated' are often the marginal lands which peasants have been forced to occupy following their displacement from land taken over by intensive, export-oriented agriculture, mining operations and so forth (Hildyard *et al.*1997).

As the case study from Bastar illustrates, taxation, the need to 'satisfy' government officials with bribes and fees, schooling, labour-saving technology, new fashions and consumerism have generated a demand for cash without the corresponding growth of a market for traditional produce (see also Colchester 1992). In other words, in situations where governments leave local people empty-handed – legally, technically, financially and politically – one can expect that sooner or later they will yield to outside forces beyond their control. With no other options before them, sooner or later they are likely to succumb to the pressures of logging firms and other commercial interests, and to lose their resources or trade them for very meagre and short-term returns.

Precisely because their own survival and cultural values are at risk, local forest-dependent communities have the strongest motivation to check the influx of illegal loggers, miners, poachers and colonists. The case studies also contribute overwhelming evidence that official efforts to restore and manage forest environments are often non-existent or both costly and ineffective. Local people – unlike the staff of government departments, international agencies or corporations – have an immediate and long-term stake in defending and evolving practices that conserve some level of biodiversity and self-reliance (Hildyard *et al.* 1997).

The majority of the case studies describe the watchdog role of local communities in response to external pressures on their forests. Yet in only a few instances do the studies mention government acceptance of the importance of local communities in controlling external use of the forest. (In a number of other countries, however, governments have acknowledged this role very explicitly. The Colombian government, for example, has handed over 20 million hectares of forest to indigenous communities in the Amazon.) Although 'the argument of local peoples' inability is used to take and maintain control over forest lands' (Colchester 1992: 16-17), caution is necessary against a romantic view of local forest management. It is unwise to portray local forest-dependent communities as homogeneous, whether they are indigenous communities in India or wood workers and their families in Canada. Local common[5] management regimes are seldom free from 'internal inequalities (particular gender inequities), back-biting, social injustices or environmentally destructive practices' (Hildyard *et al.* 1997: 13). It must be recognised, however, that, communal grazing grounds, forests and irrigation or fishing territories are an everyday reality for the vast majority of rural people. More often than not, local forest users are bound closely to each other by mutual dependence and shared values about treatment and access to the forest and other common resources, backed by social control. As Susan George emphasises, such common property regimes are managed sustainably 'so long as group members retain the power to define the group and to manage their own resources' (George in Goldman 1998: xii).

However, these communities regularly experience 'hit and run' intrusions by outsiders – such as timber merchants, traders, poachers and corrupt government officials – who roam the forest in search of quick profit. They also witness the conversion of forest by forest departments, companies or migrant colonists to establish monocultural plantations of teak, oil palm and other marketable species. These and other cultural, economic and political interventions undermine local authority, norms and values, and exacerbate inequalities. At the same time, indigenous and peasant communities often perceive such outsiders with a great sense of irony and humour, conducive to feelings of their own self-worth and dignity. The Durva people from Bastar refer to the pest *Eupatorium* as *sahib lata* (*sahib* in this context = townspeople and government officers; *lata* = weed), explaining that 'it spreads just as fast and is equally useless'. And a villager from Karnataka, on India's west coast, when confronted with corruption in the Forest Department, smilingly laments: 'When the fence is eating the grass, what can one do?'

The studies emphasise that

> where communities have a long and still vital tradition of community management, the need for the rapid re-establishment of community control over forest land is clear. However, where such traditions have long been lost due to acculturalisation and the destruction of traditional institutions,[6] the mere transition back to communal tenure and management might also prove to be destabilising and disruptive. (Colchester 1992: 21)

Many communities thus face the challenge to reassert values and to develop new methods to administer their forest lands.

Common Problems

Most often the causes of deforestation lie outside the forest and beyond the domain of the community, the district or even the nation-state. As Jeffrey Sayer notes: 'The globalisation of economies and the emergence of a strong transnational corporate sector results in significant shifts in the geographic location, type and intensity of forest use' (Sayer 1997). Most case studies describe how local people and NGOs must confront interventions by transnational companies that their own governments have done nothing to restrain. Thus a limited number of transnational corporations control an increasingly large share of logging, processing and marketing operations. In 1992 only 10 companies produced 27 per cent of the world's paper and paperboard (FOE-US 1997). The World Resources Institute calculates that commercial logging poses the single largest threat to the world's last remaining large tracts of undisturbed 'frontier' forests (notably in Canada, Brazil and Russia). The same researchers note that mining and energy development are a greater threat to these forests than agricultural expansion (Bryant *et al.* 1997: 15).

For a number of reasons, transnational companies play a major part in forest destruction and, consequently, in local socio-economic impoverishment. In the first place, they operate on a much larger scale than local companies, having the technological capacity and capital resources to open up remote and hitherto inaccessible tracts of forest. This initial penetration often sets in train further forest destruction by agricultural expansion (large monocultural cash crop plantations, colonist pioneer farming or cattle ranging). Second, the transnational impact is sharpened by globalisation, which enables world market demand for wood and paper products and other raw materials to outweigh local peoples' needs and forest conservation in determining the fate of forests. Third, foreign companies tend to take profits from forest exploitation out of the host country, instead of letting such profits benefit local people and the host economy through taxation or reinvestment (FOE-US 1997). Finally, many transnational companies show no interest in the future of the forest and allow the capital equipment of the industry (roads, mills, etc.) to deteriorate once the timber or mineral resources are exhausted. The company moves on to other regions, leaving local populations to make what they can of a devastated environment.

Privatisation of biodiversity

Some case studies, in particular the study from Brazil, also refer to the ongoing privatisation of the world's food and medicinal raw materials, notably by the agri-business and pharmaceutical industries. These industries constitute a less visible but increasingly strong lobby which monopolises – both legally and technically – an expanding share of the planet's cultural and natural domain, mainly through intellectual property protection, including patents.[7] Whereas these property systems reward human ingenuity, they ignore nature's intrinsic values and the knowledge and (informal) contribution of indigenous peoples and farmers to the maintenance and development of genetic diversity through generations of use and observation, cultivation and husbandry (Glowka *et al.* 1994). More vulnerable than the ecosystem itself, it now seems clear, is the accumulated knowledge of forest ecology held by forest-dependent peoples (Denslow and Padoch 1988). More and more forest-dependent communities are experiencing interference by outside commercial forces in their local-indigenous systems of knowledge of, access to and control over forest resources. Governments should respond urgently to the need to acknowledge, protect and reward the traditional knowledge of forest-dependent peoples, in the cause of the latter's economic and cultural survival and the interest of forest conservation.

Exporting forestry science

Until recently, global concern about deforestation has focused on the tropics and virtually excluded temperate and boreal forest issues. The case studies from Canada and the United States call into question the widely accepted belief that forestry practices in the industrialised countries are sustainable. Hence, they also question the theoretical foundations of both logging and tree-planting operations in tropical countries, which for the most part are based on temperate forestry principles (Danaiya Usher 1992). The case studies emphasise that the prevailing monetary-economic bias in conventional scientific forest resource management is in conflict with the objective of cultural and ecological diversity. Such a bias is a denial of the fact that many people are dependent for their well-being on non-monetary, ecological and socio-cultural conditions. Forest-dependent communities, which once enjoyed the comparative advantage of their skills and knowledge of a rich ecosystem, lose their culture and get pushed to the margins of society once the forest is destroyed or access to it is denied to them.

On the role of governments

These developments occur at a time when governments are being encouraged to scale down and to deregulate in order to attract foreign investment (Sayer 1997). Moreover, the case studies illustrate that some transnational companies take advantage of the political vacuum prevailing in host countries weakened by civil war, corruption or state repression.[8] One reason is that economic liberalisation is often not accompanied by political liberalisation. Whilst industry enters the hinterland in search of biomass, minerals, cheap electricity and land, local economically disadvantaged groups generally lack the formal and legal support to claim and protect their access to natural resources.

Not surprisingly, forests are unable to attract the scale of investment and attention which governments spend, for example, on roads, energy generation or the aviation industry. The situation in India illustrates this point well. While subsequent five-year plans repeat the ambitious promise to cover 33 per cent of India's land base in forest, the Forest Department is allotted less than 1 per cent of the state budget.[9] The sad irony is that the country's forest cover has dwindled to an estimated 12 per cent, a figure that is declining every year. Foresters and politicians seem to share the view that trees bring no electoral gains (Wolvekamp 1989). Like their peer organisations in most other countries in the South and in the North, the Indian Forest Service is by and large too preoccupied with generating revenue for public and private gain to forge an alliance with the tens of millions of villagers for whom the forest is their basis of survival, or to make the social need for forest protection a political issue.

The case studies thus confirm that national governments play a major role in the creation of these problems. In addition to legal shortcomings, governments are poor performers when it comes to auditing and controlling natural resource use. In many cases corruption permeates all levels of government involvement in forest management and land-use planning. The case studies question the view of many governments that forestry can generate revenues and raw material to trigger national economic development (taking monetary-economic performance as the main benchmark). These concerns were summed up long ago by Jack Westoby, former head of the forestry department of the Food and Agriculture Organisation (FAO):

> The growing interest in forestry projects had little to do with the idea that forestry and forest industries have a significant and many-sided contribution to make to overall economic and social development.... Of the new revenues generated, woefully little has been ploughed back into forestry, and the much more important role which forestry could play in supporting agriculture and raising rural welfare has been either badly neglected or completely ignored. (Westoby 1989)

Westoby spoke these words nearly 20 years ago during the Eighth Forestry Congress. Had he prepared his speech today, he might have dropped the distinction he made then between developing and industrialised countries, since in most respects his speech applies equally well to the state of affairs in many Northern countries.

Forests are under-appreciated, both for their immeasurable social and environmental services to society and for their intrinsic value. Case studies from North and South demonstrate how governments legitimise centralised large-scale forest management and intensive commercial exploitation, citing the need to protect jobs and revenues in the forest industry. Various case studies emphasise, on the contrary, that millions of people lose their jobs or sources of livelihood when access to forest sources is denied to them or as a result of ongoing mechanisation and the depletion of forest resources. Between 1990 and 1992, for example, Canada's forest industry eliminated 62,600 jobs, shedding some 28 per cent of the direct workforce (Carrere and Lohmann 1996). Notably in countries like Malaysia, Sweden, Canada and the United States, a convenient way of drawing the public's attention away from these facts, and of redirecting its concerns and anger, has been to blame environmentalists or local-indigenous communities for denying the industry access to much-needed wood resources, and in this way to hold them responsible for job losses in this sector.[10]

Economic liberalisation

The great present need for national and international regulation of investments by transnational companies is made more glaring by international agreements evolved over the last decade which are designed to facilitate 'free trade' – most manifestly represented by the emergence of the World Trade Organisation and the recent negotiation of a new Multilateral Agreement on Investment (MAI). These agreements, which have been promoted most ardently by OECD countries, curtail the freedom of nation-states to regulate foreign investment and corporate conduct and to protect vital social, cultural and environmental interests. For example, if the present draft of the MAI is approved, it is likely to overrule national as well as international environmental legislation (such as the Convention on Biodiversity) as well as the much weaker agreements which deal with human rights, minorities and indigenous peoples (such as the International Labour Organisation's Convention No. 169 and the draft Universal Declaration on the Rights of Indigenous Peoples). The MAI will prevent national governments imposing specific socio-environmental conditions on foreign investors. National governments are also restrained from reserving forest land or other national resources for local economic use, since foreign companies are given equal rights to bid for concessions. Whilst sustainable forest management demands long-term planning, the MAI forces governments to accept the immediate and unhindered withdrawal of foreign investments and profits. Transnational companies can sue national governments and demand compensation for any reduction in value of their investments as a result of social or environmental restrictions imposed by the host country's government. As a publication by Friends of the Earth–US notes: 'The MAI will throw up barriers to the types of policies needed to reverse deforestation' (FOE–US 1997).

Global master plans

Over-exploitation or destruction of forest is enhanced in countries, notably in the South and in Central and Eastern Europe, which have to deal with large foreign debts, economic depression or a process of economic transition. Their governments negotiate with multilateral financial and trade institutions avenues to open up and adjust their economies. This process bears directly on domestic land and forest policies. Many governments are advised to 'rationalise' the forestry sector. As Jack Westoby has emphasised, however, such advice generally serves foreign industry and trade interests rather than the health of forests and local or national economies.

The international financing agencies knew what foreign investors wanted, and the multilateral and bilateral agencies fell in line. They helped the under-developed countries to bear the expense and drudgery of resource data col-lection, thereby relieving potential investors of these tasks and charges. Because nearly all the forest and forestry industry development which has taken place in the underdeveloped world in the last decades has been externally oriented, aiming at satisfying the rocketing demands of the rich, industrialised nations, the basic forest products needs of the peoples of the underdeveloped world are further from being satisfied than ever. (Westoby 1989)

In the face of problems of forest loss and other environmental threats, the preferred response of many heads of industry, government agencies and multilateral institutions lies in increasingly global forms of management (see also Goldman 1998). As Hildyard *et al.* note, 'if one accepts current patterns of economic development and the institutions and premises on which they rely, the logic of "global environmental management" is impeccable' (Hildyard *et al.* 1997: 5). Sustaining this process through damage control requires an equivalent level of top-down surveillance and intervention. The physical environment becomes a terrain to be reordered, zoned, parcelled up, while people are removed or cajoled into 'collabora-tion' according to some preconceived master plan (Hildyard *et al.* 1997: 5). Through channels of aid and trade, funds are made available under the banners of development and environmental restoration (CO_2 sequestration, for example). Yet such programmes often affect forests and forest-dependent people adversely. As a number of case studies illustrate, often such funds are used to invade the countryside with infrastructural works, industrial zones or monocultural plantations.

Legal biases against forest-dependent and local people

The case studies emphasise that national laws deny millions of people access to natural resources, while most of the land is claimed by the state or engrossed by a small political and economic elite. As human and environ-mental rights lawyer Owen Lynch writes: 'National laws concerning the use and management of forest resources in at least six Asian countries (Indonesia, Thailand, the Philippines, India, Nepal, and Sri Lanka), for example, have actually become more hostile toward local people and communities than was the case during the colonial era' (Lynch 1997: 22). National laws and the way they are implemented often remain an obstacle to sustainable forest manage-ment. In many instances they reflect a lack of civil freedom to express dissent-ing opinions and the state's repression of other essential human rights. Lynch shares a conclusion reached by the case studies, namely that 'community-based forest management systems and user rights derive their legitimacy and strength from the community in which they operate, rather than from the nation-state in which they are located' (Lynch 1997: 24).

Various case studies describe how the centralisation of forest management weakened – or even abolished – local management institutions. This is most tangible where land tenure is concerned. 'Tenure systems are complex and specify under what circumstances and to what extent certain resources are available to individuals and communities to inhabit, to harvest, to inherit, to hunt and gather on, etc.', writes Lynch. Most case studies report, however, that governments deny the recognition of community-based rights over forest. Whereas in some cases the government grants certain tenurial rights, 'they are vulnerable to arbitrary cancellation' (Lynch, 1997: 26) and as such discourage local people from investing in careful, long-term use and management.

Within a context of conflict, the case studies confirm that security of land rights and user rights is the basis of forest preservation and the well-being of local forest-dependent people – especially so under conditions of external pressure. This requires awareness of their legal rights in local communities in order to defend themselves in the context of national and increasingly also international law. Better understanding of legal rights and duties also offers increased opportunities for interacting with policy makers, for example with regard to forest and land-use planning. NGOs, concerned lawyers and professional consultants often provide crucial support to bridge the gap between local aspirations and the formal language used by governments.

Plantations

Industrial plantation programmes are probably the most popular and best-sold global environmental solutions, since, it is claimed, they 'counter the greenhouse effect either by serving as carbon sinks, or by alleviating pressure on native forests, helping to preserve them as carbon depots' (Shell International and World Wildlife Fund 1996: 10). Although this claim has little substance, 'it has enough superficial plausibility to distract uninformed audiences from the more interesting topic of how to find alternatives to a system whose logic dictates a never-ending spiral in which ever-greater carbon emissions necessitate an ever more desperate search for carbon sinks' (Carrere and Lohmann 1996: 10). As some of the case studies emphasise, plantations, instead of relieving pressure on existing natural forest, add considerably to deforestation since much forest is cleared to make space for monocultural tree plantations (such as teak, gmelina, eucalyptus and poplar). World-wide, logging and plantation development go hand in hand. The logging of natural forest often provides the necessary funding for the establishment of industrial tree plantations. The plantation industry and its investors also fail to disclose that tree plantations offer only a fraction of the carbon sequestration potential of natural forests. Moreover, since these plantations are grown in short rotation cycles – and the wood is processed in short-lived products such as paper – they perform this role only temporarily.

Carrere and Lohmann give a clear definition of what a commercial plantation is and what it is not:

> Plantations, like forests, are full of trees. But the two are usually radically different. A forest is a complex, self-generating system, encompassing soil, water, micro-climate, energy, and a wide variety of plants and animals in mutual relation. A commercial plantation, on the other hand, is a cultivated area whose species and structure have been simplified dramatically to produce only a few goods, whether lumber, fuel, resin, oil or fruit. A plantation's trees, unlike those of a forest, tend to be of a small range of species and ages, and require intensive and continuing human intervention. (Carrere and Lohmann 1997: 3)

Carrere and Lohmann contrast such industrial plantations with 'attempts to plant trees in ways responsive to a wide variety of interlocked local concerns. In some agroforestry systems,[11] for example, a diversity of trees are chosen and planted to provide protection, shade and food for livestock, fruit and wood for humans, and protection, nutrients and water for crops, thus helping to keep production diverse and in harmony with local landscapes and needs' (Carrere and Lohmann, 1997: 10).

The case studies explicitly confirm that major causes of forest destruction are intersectoral in nature. Some case studies refer to the conversion of forest to other non-forest uses – for example, monocultural cash crop plantations such as citrus fruit and oil palm[12] – and they record the displacement of occupants of forest, farmland and communal grazing land as a consequence. Nevertheless, governments and institutions like the World Bank and the International Monetary Fund (IMF) continue to promote large-scale cash crop plantations, such as oil palm, as foreign exchange earners. This is probably best illustrated by the IMF's recent structural adjustment package for Indonesia. Notwithstanding the fact that the development of oil palm plantations is the single largest cause of forest fires in Indonesia, and despite increasing public concern about the scale of this social and environmental tragedy, the IMF explicitly pushes for the expansion of the oil palm sector in this country. Since the announcement of Indonesia's agreement with the IMF in January 1998, it is reported that plantation companies have continued to move into and seize the forested territories of indigenous peoples and started clearfelling and burning.

The interface with agriculture

When it comes to explaining the occurrence of high rates of deforestation in the tropics, landless farmers and traditional shifting cultivators are often scapegoated by governments, by representatives of the logging industry, and even by scientists and the media. The case studies bring out four important angles for viewing the interface between forests and agriculture.

First, unauthorised conversion of forest land to agricultural use indeed bears responsibility for much deforestation, as the case study from the Democratic Republic of Congo illustrates. Case studies from, for example, Paraguay and India remind us of the role of politicians who endeavour to satisfy the electorate by endorsing the encroachment on public forest land, bearing in mind the adage that 'trees don't vote for you' (Wolvekamp 1989). The case studies question, however, the common practice of blaming local indigenous people and peasants in order to veil forest destruction due to government-sanctioned logging and cash crop plantations. As Jeffrey Sayer emphasises, official government-registered programmes of forest conversion for migrant agriculture and large-scale commercial cash crop plantations are a far greater cause of deforestation (Sayer 1997). Hence there is a need for accurate maps and data on actual land use and planning to better inform public debate.

Second, as more and more fertile land is claimed for growing export crops, the rural population is pushed to marginal lands and forced to clear the remaining forest cover in order to eke out a living. This is a strong argument in favour of better land distribution and land-use planning: 'ecological stress on forest land would be better relieved by reclaiming "high potential areas" for peasant agriculture' (Hildyard *et al.* 1997: 5).

Third, as the case study from Bastar vividly illustrates, the great confusion surrounding terms like 'shifting cultivation' and 'forest' is the cause of 'facile and unsophisticated assessments of the effects of shifting cultivation by governments and influential international organisations' (Sunderlin 1997: 11). William Sunderlin notes that 'A major positive development in the recent debate on the role of shifting cultivation has been that influential analysts of the forest situation are no longer willing to accept at face value the claim that shifting cultivation is uniformly bad for forest conservation and management' (Sunderlin 1997: 8). There is, for example, growing recognition of the need to distinguish, roughly, between 'shifting cultivation' and 'forest pioneer' farming systems.[13] Those who argue that shifting cultivation is a threat to forests are actually referring to forest-pioneer farming or short-fallow shifting cultivation (Sunderlin 1997: 8).

Fourth, the case studies from Cameroon and Bastar demonstrate that modern 'forestry science separated forest management strictly from agriculture, and focussed almost exclusively on production of uniform quantities and qualities of timber' (Carrere and Lohmann 1997: 10), while 'conventional forestry is also firmly based on legal notions which diverge strongly from local peoples' own frame of thinking' (Brocklesby and Ambrose-Oji 1997: 17). Cameroon's forestry law illustrates this point well, defining a forest as 'any land covered by vegetation with a predominance of trees, shrubs and other species capable of providing products other than agricultural products' (Forestry, Wildlife and Fisheries Regulations, Law No. 94/01 of 20 January 1994, in Brocklesby and Ambrose-Oji 1997: 17). This definition has its roots in a policy dichotomy, found throughout the world, between agriculture and forest. 'Whilst this (definition) may support large plantations and production forests, it ignores the farm/forest interface' (Brocklesby and Ambrose-Oji 1997: 18). This means that in Cameroon, as in so many other countries, 'legally agreed management plans (which form the basis upon which a community can establish a community forest) cannot by law include regulations governing shifting agriculture plots and fallow use, since these practices are not recognised' (Brocklesby and Ambrose-Oji 1997:18). The case studies offer concrete examples of viable symbiosis between agriculture and forest conservation through the management of non-timber forest products,[14] agro-forestry systems such as Analog Forestry,[15] and other approaches.

Transport

The promotion of domestic and cross-border traffic, which is part and parcel of the drive towards regional economic integration – the European Union, for example, or the North American Free Trade Agreement (NAFTA) – is another major cause of forest destruction. Transnational corporation lobby groups the world over are successful in persuading governments and multilateral agencies to spend huge amounts of taxpayers' money on infrastructural projects – as when the European Round Table of Industrialists prevailed on the European Commission to adopt its proposals for the Trans-European Network, the largest transport infrastructure plan in history. The Commission presented its plan for the development in Eastern Europe of some 38,000 kilometres of new motorways, high speed railways, new harbours and airports: 90 per cent of a total cost of US$100 billion is to be paid by the Eastern European countries themselves (*De Volkskrant*, 25 June 1998; see also Corporate Europe Observatory 1997). A potential disaster in the making is the Chad–Cameroon Petroleum Development and Pipeline project, primarily sponsored by an oil consortium consisting of Exxon, Shell and Elf Aquitane. The project plans to develop three oil fields in southern Chad using an export system including a 1,050-kilometre pipeline, most of which passes through Cameroon, and an off-shore loading facility for crude oil on Cameroon's densely forested southern coast. The World Bank is considering loans to the governments of Chad and Cameroon to help finance their respective portions of equity (amounting to about 15 per cent) in two pipeline companies in which the majority share (80 per cent or more) will be owned by the oil consortium. There is growing concern that the project will lead to escalating civil violence, notably in southern Chad, and bring social and environmental destruction

– due to oil spills, for example – to the people living downstream in Chad and to Pygmy communities near Kribi on Cameroon's south coast. Other examples are the Trans-Amazonian highway – built to give Asian markets, particularly, access to the Amazon's timber and minerals – and the Hidrovia river canalisation project of the Mercosur countries, which will dry out Brazil's Pantanal (earth's largest wetland, containing its highest diversity of mammals). These and other new transportation routes will open up some of the world's last frontier forest areas to logging, cattle-ranching, mining, industry and poaching – and will expose local populations to the inevitable violence of these incursions and to increased pollution (Goldsmith 1997).

The price of opposition
The case studies also show that the state-sanctioned, 'legal' usurpation of natural resources to the detriment of forest-dependent rural people often goes uninterrupted because many local communities do not fully understand their legal rights and options (Lynch 1997). Moreover, many local constituencies wishing to voice their interests or seek legal redress face major practical and cultural obstacles – logistic, financial, technical, linguistic – when approaching decision makers or the judiciary in the national and provincial capitals.

In this respect, the case studies also point to the ambivalence of governments, industry and international development agencies towards local people and their organisations. On one hand, these institutions tend to ignore or deny the important skills, knowledge and vision of local people and their supporters. On the other hand, often local people and NGOs are persuaded to 'participate' in order to lend legitimacy to the process of developing a project – a hydroelectric dam, a social forestry project or a mining site. As the case studies illustrate, genuine public participation has often been the result of local mobilisation against an unwanted development activity. It often goes unnoticed, however, that local communities and supporting NGOs, from both South and North,[16] have mounted their opposition in the face of violence, land deprivation and recurrent intimidation (Lynch 1997).

During the period covered by the case studies, no fewer than seven participating organisations faced severe hardship due to civil war, criminal violence, legal battles and conflicts with government authorities. Five staff members were killed, property was destroyed and people were arrested.

Conclusions and Recommendations

Opportunities and challenges
Looking at the political and economic root causes of deforestation, there is reason to wonder whether there is scope to direct society, and in particular

the powers that be, in a more sustainable direction. But a choice for apathy or cynicism would mean abandoning those who work at the 'front line' and attempt to change things for the better. The case studies reveal heartening responses by local peoples to problems encountered. Often their vigilance has been rewarded by new chances for forest conservation and local benefits from forest management.

The failure of most governments to recognise the role of local forest management has not necessarily terminated local management of and tenure over forest resources. As Owen Lynch observes, 'Despite expansive claims of ownership, many national governments exercise relatively little control over large areas of forest ... [since] few governments have the staff needed, and the commitment, to survey, patrol and effectively manage vast areas classified as state-owned' (Lynch 1997: 24). Under such circumstances, many governments merely tolerate the presence of local peoples in the forest, while their systems of natural resource management are often branded backward and unsustainable, or as encroachment.

More positively, the case studies also refer to a growing range of initiatives and opportunities to foster collaboration between local people, state authorities and other parties in support of local, sustainable forest management. Local forest-dependent communities and their supporters face a dual challenge: first, they must counter external forces which threaten local access to forest; second, they must prove that their local system of forest management is potentially viable.

We have listed some general lessons and suggestions for political and practical action that emerge from the empirical findings of the case studies. Inevitably, measures are required in different fields – including the economy, the environment, culture, governance and law – and at various levels.

The role of research

Local forest management practices often remain invisible, only coming to light when there is a clash of interests within or between local communities and the outside world. Most case studies narrate what can be summarised as local attempts to manage a 'natural resource conflict situation' (Daniels and Walker 1997). The studies describe strategies and experiences of local people and their allies in attempting to change political conditions, opening space for local forest management and improving their position on the ground. The participating organisations used the case studies compiled in this volume as tools for developing self-assessment, policy dialogue and concrete management. Action research, undertaken by local people or in partnership with them, is essential in breaking the cycle of isolation and anonymity.[17]

Scientific research, notably environmental analysis and information, is often indispensable for land-use planning and policy formulation. To ensure that land-use policies take into account the aspirations and needs of people inhabiting such areas, much more insight into local forest use and management is required. Research can be instrumental in gaining recognition for the knowledge and perceptions of local people, in disclosing conflicts of interest and in negotiating solutions. In doing so, it may help prepare common ground where local people and outsiders – such as forest department personnel or conservationists – can meet, negotiate and even start to collaborate.[18]

Considering the complex linkage between forest and agriculture in the tropics, it is important that future research offers more insight into the constraints imposed on smallholder farmers by other competing land uses (plantations or mining, for example) and market distortions, such as the dumping of foodgrains in Southern countries by the European Union, as possible obstacles to more symbiotic relationships between local farming systems and the forest environment. Another research challenge is the modification and development of traditional shifting cultivation, thus allowing it to play a positive role in forest conservation. Priority attention also needs to be given to alternative farm systems and to non-farming sources of subsistence and income that can alleviate pressure from pioneer farming on forests (Sunderlin 1997).

More research should be undertaken by local people themselves or in partnership with them. There is still a vast storehouse of local-indigenous, forest-related culture and experience which needs to be documented systematically. This will create wider appreciation of the value of forests to local societies, in particular to men, women and children at the household level, and will generate greater recognition of the potential constraints and needs of local forest management systems.

Redressing legal biases

Security of local land rights and user rights is the basis of forest preservation and the well-being of forest-dependent people. It calls both for the recognition of customary land titles and for greater collaboration between governments and local people, who should be entrusted with the management of public (forest) lands on condition of sustainable use. At the same time, legal arrangements need to be made to achieve genuine land reform, as an alternative to the politically more convenient practice of handing out public forest land for agricultural purposes. Reform at both national and international levels is required to address the legal bias against forest protection and the customs and rights of local people. Legislators and legal professionals should assume responsibility for establishing a tradition of public interest environmental law, through training and the adjustment of public and civil law. Moreover, NGOs, lawyers and legal experts have a responsibility to popularise national and international law, important tools with which citizens can keep their governments accountable and achieve recognition and protection of their human rights and of environmental values. It is essential that more attention and support should be given to initiatives which explore and propagate existing legal provisions on community forestry and the recognition and restoration of ancestral territorial rights. Governments are urged to endorse ILO Convention No. 169[19] in recognition of the rights of indigenous peoples. However, we should not look only to legislatures, courts and other governmental institutions to introduce such legal reform. On the contrary, as Owen Lynch concludes, 'We are all law makers, and it behoves us to work together to develop better legal strategies and tools ...' (Lynch 1997: 28).

The case studies make a strong plea that politicians and civil servants, notably those from OECD countries, should start realising the adverse social, ecological and economic implications of ongoing privatisation and monopolisation of food and medicinal raw materials, and that laws should be passed which put a halt to the 'selling out' of biodiversity and indigenous local knowledge.[20] It is recommended that governments declare biodiversity to be part of the public domain in each country in order to stop further privatisation.

Enhancing self-organisation

The case studies show that efforts to protect or repair the interests of local people and their forest environment have invariably started with a great investment of time and commitment to foster unity and a common direction within the community. Actions to prevent outsiders from exploiting local forest wealth, for example, have started with efforts to strengthen the local social fabric and legal position. Systems of decision making, local knowledge, the improvement of local management practices and enhanced bargaining power are vital parts of successful local responses to external pressures.

The current economic crisis in Indonesia shows that millions of rural people, no longer able to obtain their basic needs from the market, cannot fall back upon traditional subsistence practices because much of the natural environment has been destroyed by government-sponsored timber estates, oil palm plantations and forest fires. To avoid such risks, governments should seek to ensure that the power and means to achieve economic survival and development are located as close to the people as possible. Greater economic self-sufficiency and self-determination should be supported, without the assumption that local communities can supply all their needs (Daly and Cobb 1989). As Hildyard *et al.* conclude, 'only when all those that have to live with a decision have a voice in making that decision can the checks and balances on power that are so critical to the working of the commons be ensured' (Hildyard *et al.* 1997). Governments should recognise that most forests on which communities rely – in the South as well as in the North – must be considered as commons. Hence, governments, donor agencies and NGOs need to support the building of open, accountable institutions that consolidate or restore the authority of commons regimes.

'The mapping of their land[21] and resource use helps local communities to protect their land from outside incursions and thereby lessens demographic pressures on fragile ecosystems' (Lynch, 1997: 15). It thus constitutes a vital element within the process of self-organisation and articulation of claims. If communities come together to map their lands and discuss regional development, local people can acquire a broader perspective of the extractive pressures in the region, and a sense of how these will affect them. It is often essential to include representatives of neighbouring communities – and, if opportune, other forest users – in their discussions, in order to avoid, or mitigate, conflicts of interest. Moreover, the whole mapping process and its legal underpinnings may encourage collaboration between local communities and conservation authorities or other management institutions (Lynch 1997). It is recommended that NGOs, donors, governments and scientists support the process of mapping by local communities as a tool for information sharing, negotiation and land-use management.[22]

To advance self-organisation further, policy makers and forest authorities should support full recognition of local use and management of non-timber forest products (NTFPs). These products are of significant importance in rural areas, especially among disadvantaged groups such as the landless poor who have access to few resources. Furthermore, NTFPs represent a direct and potentially positive connection between forest conservation and forest use. If farming communities living on the fringes of the forest also derive value from the sustainable exploitation of NTFPs, this offers them an incentive to protect the forest (de Beer and MacDermott 1997). Perhaps most significantly, NTFPs can help create or restore a positive interface between agriculture and forest conservation.[23] Donors, government agencies, NGOs and scientists must assist the management of NTFPs by local people through legal provisions and support for local capacity building. More particular assistance is required in the following fields:

1 Removing legal obstacles which hinder local people who seek to manage and benefit from NTFPs;
2 Technical and institutional strengthening (in areas such as administration and marketing);
3 Sustainable extraction of NTFPs from the wild and cultivation of NTFPs when appropriate;

4 Strengthening the position of women, notably those who belong to marginal groups;

5 Information sharing and capacity building through exchange of experiences among local communities.[24]

Alliances for the future

Local forest-dependent communities and their support organisations often experience a vicious circle of isolation and the inaccessibility of contacts, information, financial means, recognition and political support. Unless this circle is broken, local forest management practices will not have an opportunity to prove their potential as a more sustainable alternative to dominant systems of forest exploitation. This is an area in which donors, NGOs, consultants and governments have most to offer in terms of redistribution and the regulation of access to natural resources. They should aim at enhancing possibilities for marginal groups to claim and protect their access to such resources. This requires a new sensitivity to the needs and priorities of forest-dependent people and their local resource management systems. The case studies offer convincing experiences of how such collaboration can lead to an increased capacity to manage conflicts over forest. Whereas recognition of tenurial rights is essential, in itself it is not sufficient. Governments and donors need to ensure that the provision of technical assistance, along with credit and health programmes, responds to the needs and perceptions of local communities. Notably in the interface between agriculture and forest, local people, NGOs, scientists, governments and donors face the challenge of supporting approaches which balance the objectives of food security, economic welfare, self-determination and conservation. Faced with deteriorating environments and poverty, local people require an opportunity to develop internal coherence based on alternative sources of income and livelihood if they are to prevent forest destruction.

Those who wish to collaborate with local stakeholders should also be prepared to make a long-term commitment to building trust and partnership. This plunges one into a reality which differs from the reality of those officials, bankers and consultants who keep their distance from the field – yet it is often these more distant groups which make far-reaching decisions about the future of forests and forest people, without witnessing the consequences. Development agencies and other external agents thus have to make clear choices when it comes to collaboration (Hildyard *et al.* 1997). As Larry Lohmann argues, 'Blaming client governments or their departments when a project stifles participation of local people in forest management, for example, should have no place in agencies that are committed to fostering genuine participation and local control. It should be the responsibility of agency staff to evaluate in advance whether or not a partner government is likely to support local participation and not to become involved if this evaluation is negative' (Lohmann 1993, in Hildyard *et al.* 1997: 24). Donors, scientists and governments should link up with local initiatives and give primacy to the needs and political demands of marginalised and oppressed groups. This may require them to take measures that actively disempower dominant groups – for example, by promoting agrarian reform and by enhancing the position of women (Hildyard *et al.* 1997).

Additional mechanisms of flexible funding, especially small grants schemes, need to be developed in support of the work of local communities, NGOs and individuals in the field of forest preservation and management. Great emphasis should be placed on making such funds available for activities in the temperate and boreal forest regions as well, thus including the OECD countries. Donors should give priority attention to strengthening the position of politically marginalised groups. Bilateral donors and multilateral financiers are urged to make community forestry and non-displacement of local people conditional upon their funding, since this may help prevent external funding adding to a downward spiral of poverty and environmental degradation. Moreover, NGOs and donors should make their own participation in programmes led by international agencies, the corporate sector or governments[25] dependent upon the degree to which such initiatives embody a genuine commitment to structural change and address the political demands of marginal groups.

Addressing the root causes of deforestation

The case studies point out that any attempt at consolidating or restoring local systems of forest management requires, in the first place, that underlying causes of forest destruction should be addressed. These causes are to be found outside rather than inside the forest. The studies emphasise the cross-sectoral linkages: for example, the politics of energy, agriculture and transport bear directly on forests and forest-dependent economies and cultures. The participating organisations from Costa Rica, the Philippines and Ghana also explain that IMF and World Bank structural adjustments programmes (SAPs) accelerated forest destruction in their respective countries.

The prevention of poverty and further environmental destruction demands, first of all, that societies in the West and in emerging economies in the South abandon increasingly unsustainable levels of consumption and production. There is the challenge to design and adopt socially and ecologically benign avenues towards needs satisfaction and fulfilment. The dominant perception, it appears, is that forests, and nature in general, are a mere subsystem of the economy, instead of vice versa. The case studies describe how forests and the survival of forest-dependent people are sacrificed to what we wish to call a 'free rider economy'. There is, in the words of Michael Goldman, 'The problem ... that when wealth is defined in purely economic/quantitative terms, most social labour, ecological processes and cultural world views become devalued ... [and] remain outside an economic calculus. That is, without the unpaid labour from the commons, the household and the community, and without tapping ecological processes, there could not be any surplus-value production for capitalist industries' (Goldman, 1998:16).

Therefore, governments, donors and international economic institutions (the IMF, the European Union and the OECD, for example) need to prepare an answer to the fact that the current wave of unchecked economic liberalisation is rapidly undermining the ecological and cultural basis of livelihood of millions of vulnerable groups and of the economy in general. This calls, first of all, for fiscal reforms, adapted trade agreements and formal investment policies and regulations. Politicians, scientists, citizen groups and civil servants, from both North and South, are encouraged to collaborate in demanding a public debate on the proposed Multilateral Agreement on Investment and related negotiations.

The primary goal of forest management and reforestation programmes should be to enable forests to perform their many vital ecological functions and to benefit people who depend on forests as a source of income and for their shelter, food, firewood, fodder, medicine and other basic needs.This calls, for example, for governments and donors to choose enhanced natural regeneration of secondary forest and agro-forestry systems as options preferred to monocultural industrial plantations. Likewise, greater priority should be given to maintaining the carbon store in existing natural and old-growth forests, a course of action which in the end is of greater social and ecological benefit to society than the introduction of plantations.

The case studies confirm that commercialisation of forest resources should only be pursued if, and to the extent that, this does not compromise the well-being of people and ecosystem integrity (Colfer *et al.* 1995). Donors should assist local-indigenous communities, NGOs and governments in the South with technical and financial support to prevent the commoditisation and expropriation of biodiversity and traditional knowledge.

Commercial enterprises (mining and logging companies, for example) which do not accept the primacy of local communities' needs and which do not respect them as their equal partners in development and conservation activities, should not be permitted to operate in such areas. This calls for more transparency about the aims, motives and methods of forest use, so as to enable the general public to increase their participation in the control and protection of the nation's forest wealth. Hence, investors and companies should face closer scrutiny than before from governments, shareholders, NGOs, the media and – increasingly – their own staff. North–South collaboration and information sharing is essential in ensuring that commercial activities in one part of the world help to determine a company's reputation – and profitability – in countries and regions thousands of miles away.

NGOs, donors and governments are urged to explore the possibilities of establishing a platform – a tribunal or an ombudsman, perhaps – where affected citizens and other concerned parties (such as NGOs and scientists) can seek impartial judgement, protection and means of redress. An independent ombudsman, both at the national and international levels, deserves special attention as a last resort for facilitating the access of citizens to sources of justice, public opinion and arbitration.[26]

Governments and international institutions like the International Chamber of Commerce should address with priority the problem of 'free rider' companies which continue to enjoy the benefits of market access without adhering to international standards.[27] Investors and corporations need to be subjected to alert and critical observation backed by visible and independent monitoring; they should also be exposed to positive incentives. The object of these measures is to enable consumers, investors and the public at large to recognise and distinguish between good corporate performances and those businesses which fall short of the standards set.[28] Donors should respond to the growing scope for independent public interest organisations which inform the public in general, and market partners such as consumers and investors in particular, about corporate performance in forest management. In addition, NGOs and governments should explore the possibility of introducing the principle of 'immobilising capital'.[29] This principle makes the issuing of licences, concessions or permissions to exploit natural resources (by mining or logging, for example) by (foreign) companies conditional on the company's lodging a suitable security.[30]

More generally, local forest users should be encouraged and enabled to present their own experiences and priorities in order to inform public opinion, since 'only a well-informed and enfranchised public will be concerned enough to see the flaws in the present system and demand alternatives' (see Chapter 13, p. 203).

A potential conflict remains between local control over forest resources and over-exploitation. Local people, however, have most to lose from forest destruction and in many instances the responsibility for the long-term protection of forests rests with them. This is why the case studies make their strong plea for a continuous investment in local people: to consolidate or strengthen their ability to defend and sustain the forest for their own immediate benefit, and for society at large.

Paul Wolvekamp

NOTES

1 With thanks to Eline Meyer.

2 The boreal forests, known as the *taiga* in Russian, are one of the world's three great forest ecosystems. The *taiga* covers approximately 920 million hectares and can be seen as a green belt encircling the northern hemisphere, stretching from Alaska in the west to northern Russia in the east. The boreal forests are characterised by coniferous tree species such as spruce, pine and fir and broadleaved species such as alder, birch and poplar. (With thanks to Ann Janssen.)

3 Different forms of 'shifting cultivation' are explained in the next paragraph, under the heading 'The interface with agriculture'.

4 Thanks to Kaki Buti from Palob village, Bastar and Madhu Ramnath.

5 For example, grazing lands, village forests, fishing grounds are local commons, which are communally owned and/or used and looked after.

6 Such institutions encompass, amongst others, the regulations, norms, values, sanctions and rewards which determine leadership, division of tasks and the rights and responsibilities of all men, women and children concerning the maintenance, protection and distribution of land, water, flora and fauna, conduct *vis-à-vis* the spirits and deities, and other religious and cultural aspects of their lives.

7 An international agreement – Trade-Related Aspects of Intellectual Property Rights (TRIPs) – was signed in early 1994 as a result of the Uruguay Round of the General Agreement on Tariffs and Trade (GATT). Following extensive pressure from Organisation for Economic Cooperation and Development (OECD) countries, TRIPs introduced mechanisms to recognise, claim and enforce intellectual property rights.

8 Other noteworthy examples are to be found in countries such as Cambodia, Burma, Liberia, Indonesia and Nigeria.

9 The ministries of irrigation and power, by contrast, sometimes receive over 20 per cent of the budget.

10 The former Republican US vice-president, Dan Quayle, was a major champion of this approach in his campaign against conservation measures meant to save the remaining old-growth forests of Oregon – the region considered in the case study of the Confederated Tribes of Warm Springs.

11 A form of land use whereby the growing of trees is deliberately integrated with crops and animals on the same land management unit, either at the same time or in sequence with each other (International Centre for Research in Agroforestry, annual report, 1993).

12 Other crops which need to be mentioned are tobacco (according to Goldsmith an estimated forest area of 12,000 square kilometres is felled every year to fuel tobacco-curing barns (Goldsmith 1997), rubber, coffee and soya. And prawn cultivation for export is a major reason why about half of the world's mangrove forests have been cut down, with catastrophic consequences for local fishing communities.

13 Sometimes referred to as 'swidden agriculture' or 'slash-and-burn agriculture'. Shifting cultivators could be defined as people who practise a form of rotational agriculture with a fallow period longer than the period of cultivation, whereas forest pioneers may slash and burn existing vegetation but have the primary intention of establishing permanent or semi-permanent agricultural production. The planting of cash crops is the primary focus of attention. (J. A. Weinstock and S. Sunito, 'Review of Shifting Cultivation in Indonesia' in Sunderlin 1997: 4).

14 'The term non-timber forest products encompasses all biological materials other than timber which are extracted from forests for human use. These include foods, medicines, spices, resins, gums, latexes....' (Jenne H. de Beer and Melanie J. McDermott, *The Economic Value of Non-Timber Forest Products in Southeast Asia,* second revised edition, Netherlands Comittee of the International Union for the Conservation of Nature (IUCN), 1996).

15 See the case study by Arbofilia, Costa Rica, for an elaborate explanation of the Analog Forestry approach.

16 For example – as the participating organisations from Canada themselves experienced – the legal system in North America offers corporations wishing to break the opposition from indigenous grassroots activists and environmentalists the opportunity to use 'Strategic Lawsuits Against Public Participation' (or SLAPP suits) 'to sue them for defamation, injury, conspiracy etc., in order to bring victims to the point where they are no longer able to find the financial, emotional and mental wherewithal to sustain their defence' (Edwards 1997).

17 Examples might include formulating a forest management plan; convincing the government of the need to give a particular forest a protected status and to recognise the land rights of indigenous people; stimulating debate on legislative amendments; launching an awareness campaign; developing working relations with donor organisations and relevant experts; presenting local experiences and views to international institutions and fora; and, – last but not least – exchanging experiences with other local organisations.

18 Many local-indigenous people appreciate the idea of 'living' in the forest – which implies an inclusive, holistic, perspective – more readily than the concept of 'managing' the forest, implying a detached, exclusive perspective. A more profound understanding of local forest-use situations – their potentials, requirements and constraints – may facilitate an interaction between conventional forest management and indigenous forest-use practices. This, in turn, may contribute to a critical examination of the conventional approaches.

19 The ILO Convention affirms that 'The rights of ownership and possession of the peoples concerned over the lands which they traditionally occupy shall be recognised' (Article 14.1) and that 'The rights of the peoples concerned to the natural resources pertaining to their lands shall be specially safeguarded' (Article 15.1).

20 See, for example, the Preamble and articles 15 and 16 of the Convention on Biodiversity. Governments are encouraged to make constitutional provisions or other legal mechanisms on intellectual property rights which incorporate relevant principles of the Biodiversity Convention (in particular articles 15 and 16).

21 Indigenous peoples often prefer the term 'territory' when referring to ancestral land.

22 See a workshop report by Instituto del Bien Comun, Local Earth Observation and Center for the Support of Native Lands, Geomatics and Indigenous Territories, Hacienda San Jose, Chincha, Peru, 25-29 June 1998.

23 This once more reiterates the importance of traditional knowledge of the complexity of the forest ecology, and of NTFPs in particular as agents of seed dispersal and pollination, and as elements in natural food chains.

24 For example the NTFP Exchange Programme for Southeast Asia – a joint endeavour by Both ENDS and the Dutch consultancy firm ProFound in collaboration with the Philippines-based federation NATRIPAL (United Tribes of Palawan) with the support of the Netherlands committee of IUCN – aims at local capacity building by facilitating exposure visits, local and regional meetings and the production of a modest newsletter which compiles practical information on matters such as sustainable NTFP harvesting, marketing and land tenure.

25 Examples include 'debt for nature swaps', 'joint implementation', 'joint forest management', covenants with the industry, Global Environment Facility projects and 'green investments'.

26 The ombudsman can perform the following roles: (1) act as a watchdog; (2) give impartial judgement with reference to relevant laws and regulations, national or international jurisprudence, and generally accepted norms of good conduct; (3) offer a platform for mediation; and (4) inform public opinion.

27 These standards include, for example, the UN Declaration of Human Rights and the guidelines issued by the Forest Stewardship Council.

28 Poor performers treat social and environmental values as externalities which can be shifted to the political and economic fringes, or to following generations.

29 I am grateful to Didier Babin and colleagues of Centre de Cooperation International en Recherches Agronomiques pour le Développement (CIRAD), France, for this information.

30 When the contract runs out, the security can be claimed if the company's operations have led to damage to the environment, affected local communities adversely or injured the national treasury – by evading taxes, for example, or not paying royalties. Companies are thus subjected to the widely accepted custom that tenants renting a furnished room pay key money as a guarantee. It is suggested that the administration of securities should be dealt with by independent institutions.

BIBLIOGRAPHY

Beer, J. H. de and MacDermott, M. J. (1997) *The Economic Value of Non-Timber Forest Products in Southeast Asia*, Netherlands Committee of the International Union for the Conservation of Nature (IUCN), Amsterdam.

Brocklesby, M. A. and Ambrose-Oji, B. (1997) 'Neither the Forest nor the Farm .... Livelihoods in the Forest Zone – the Role of Shifting Agriculture on Mount Cameroon', ODI Network Paper 21D.

Bryant, D., Nielsen, D. and Tangley, L (1997) *The Last Frontier: Forests, Ecosystems and Economies on the Edge*, World Resources Institute.

Carrere, R. and Lohmann, L. (1996) *Pulping the South: Industrial Tree Plantations and the World Paper Economy*, Zed Books, London.

Chambers, R. (1983) *Rural Development: Putting the Last First*, Wiley, New York.

Colchester, M. (1992) 'Sustaining the Forests: Community-based Approaches in Southeast Asia', United Nations Research Institute for Social Development (UNRISD) research paper.

Colchester, M. (1997) 'National Sovereignty, Free Trade and Forest Peoples' Rights', Intergovernmental Forum on Forests, position paper.

Colfer, C. J. P. and Dudley, R. G. (1993) *Shifting Cultivators of Indonesia: Marauders or Managers of the Forest?* Community Forestry Case Study Series 6, FAO, Rome.

Colfer, C. J. P., in collaboration with Prabhu, R. and Wollenberger, E. (1995) *Principles, Criteria and Indicators: Applying Ockham's Razor to the People–Forestry Link*, Centre for International Forestry Research (CIFOR) Working Paper No. 8.

Corporate Europe Observatory (1997) *Europe Inc. Dangerous Liaisons between EU Institutions and Industry*, Amsterdam.

Daly, H. E. and Cobb, J. B. Jr (1989) *For the Common Good. Redirecting the Economy toward Community, the Environment and a Sustainable Future*, Beacon Press, Boston.

Danaiya Usher, A. (1992) *Taiga News*, No. 4 (December), editorial.

Daniels, S. E. and Walker, G. B. (1997) 'Rethinking Public Participation in Natural Resource Management: Concepts from Pluralism and Five Emerging Approaches', paper presented to the Workshop on Pluralism, Sustainable Forestry and Rural Development, FAO, Rome.

Denslow, J. S. and Padoch, C. (1988) *People of the Tropical Rainforest*, University of California Press.

Edwards, D. (1997) 'Old Wine, New Bottles', book review of Sharon Beder, *Global Spin: the Corporate Assault on Environmentalism* (Green Books), in *The Ecologist*, Vol. 27, No. 6 (November/December).

Falconer, J. (1991) *Nature et Fauna*, Vol. 2.

Friends of the Earth–US (FOE–US) (1997) 'The MAI and Global Deforestation', draft, October 1997.

Fisiy, C. P. (1989) 'The Death of a Myth System and Land Colonisation on the Slopes of Mount Oku – Northwest Province of Cameroon', unpublished paper.

Glowka, L., Burhenne-Guilmin, F., Synge, H. *et al.* (1994) 'A Guide to the Convention on Biological Diversity', Environmental Policy and Law Paper No. 30, International Union for the Conservation of Nature (IUCN).

Goldman, M. (ed.) (1998) *Privatizing Nature: Political Struggles for the Global Commons*, Pluto Press in association with the Transnational Institute, London.

Goldsmith, E. (1997) 'Can the Environment Survive the Global Economy?', *The Ecologist*, Vol. 27, No. 6.

Hildyard, N., Hegde, P., Wolvekamp, P. and S.T. Somasekhare Reddy (1997) 'Same Platform, Different Train: Power, Politics and Participation', paper for the Worksop on Pluralism, Sustainable Forestry and Rural Development, FAO, Rome.

Jepma, C. J. (1995) *Tropical Deforestation: a Socio-Economic Approach*, Earthscan Publications, London.

Lynch, O. (1992) 'Securing Community-based Tenurial Rights in the Tropical Forests of Asia: an Overview of Current and Prospective Strategies', briefing, World Resources Institute.

Lynch, O. (1997) 'Legal Aspects of Pluralism and Community-based Forest Management: Contrasts between and Lessons Learned from the Philippines and Indonesia', Centre for International Environmental Law, paper for the Worksop on Pluralism, Sustainable Forestry and Rural Development, FAO, Rome.

Netting, R. McC.(1997) 'Unequal Commoners and Uncommon Equity: Property and Community among Shareholder Farmers', *The Ecologist*, Vol. 27, No. 1.

Olssen, R. (ed.) (1995) *The Taiga Trade: a Report on the Production, Consumption and Trade of Boreal Wood Products,* Taiga Rescue Network.

Perpongsacharoen, W. and Lohmann, L. (1989) 'Some Thoughts on Action on the Tropical Forest Crises', position paper for NGO distribution networks.

Posey, D. A. (1996) *Traditional Resource Rights: International Instruments for Protection and Compensation for Indigenous Peoples and Local Communities*, International Union for the Conservation of Nature (IUCN).

Rietbergen, S. (ed.) (1993) *The Earthscan Reader in Tropical Forestry*, Earthscan Publications Ltd, London.

Sachs, W., Loske, R., Linz, M. *et al.* (1998) *Greening the North: a Post-industrial Blueprint for Ecology and Equity*, a study for the Wuppertal Institute for Climate, Environment and Energy, Zed Books, London.

Sayer, J. (1997) 'Changing Roles in Forest Research', keynote address at the Tropenbos seminar 'Research in Tropical Rainforests', Wageningen.

Shell International Petroleum Company and World Wildlife Fund for Nature (1996) *Tree Plantation Review*, 11 vols, London, in Carrere and Lohmann, 1996.

Shiva, Vandana (1989) *Staying Alive: Women, Ecology and Development*, Zed Books, London.

Sunderlin W. D. (1997) *Shifting Cultivation in Indonesia: Steps towards Overcoming Confusion in the Debate*, ODI Network Paper 21 B.

Weinstock, J. A. and Sunito, S. (1989) *Review of Shifting Cultivation in Indonesia*, Directorate General of Forest Utilisation, Ministry of Forestry, Government of Indonesia and FAO,, Jakarta.

Westoby, J. (1987) *The Purpose of Forests: Follies of Development*, Basil Blackwell, Oxford.

Wolvekamp P. S.(1989) 'Trees Don't Vote for You. Het Functioneren van het Karnataka State Forest Department', MSc thesis, University of Leiden.

Introduction to the Case Studies

The seventeen case studies brought together in this book have been recorded for different purposes in response to specific threats and opportunities. Most of the case studies describe the struggle of local communities and their supporters for the preservation of their forest and their rights over such areas. In this category are the three cases from India (by Nature's Beckon in Assam, by Prakruthi about Halikar Village in Karnataka, and by Madhu Ramnath in Bastar); the account by the Montanosa Research and Development Centre from the Philippines; the collaboration between Acción Ecológica and the Chachi people in Ecuador; the cases from the United States and Canada; the study about the Udege people in Far Eastern Russia; and the account of the Stramproy woodlot owners' association in the Netherlands.

Although the other case studies also focus on local struggles towards recognition of rights and forest protection, there is more emphasis on broader analysis to reach conclusions and strategic recommendations for political change. These are the case studies contributed by Vitae Civilis from Brazil, Centre pour l'Environnement et le Développement (CED) from Cameroon, Sobrevivencia from Paraguay, Arbofilia from Costa Rica, Green Earth Organisation (GEO) from Ghana, Centre d'Action pour le Développement Durable et Intégré Dans les Communautés (CADIC) from the Democratic Republic of Congo and the Polish Ecological Club.

Most of the case studies offer up-to-date basic information about the ecological and socio-economic circumstances which impact on local forest conditions and the welfare of local communities. This chapter places the case studies in a somewhat wider national and global political context to facilitate comparison and to help distinguish some dominant trends.

Asia

Four organisations from Asia participated in this survey of local forest management practices. In India, where three of the case studies were carried out, the colonial roots of forest policy are still clearly visible. The state took early steps to control the forest and its inhabitants. In 1866 the Imperial Forest Service was created and a system of scientific forest management was introduced with the aim of maximising timber yields and revenue. The Indian Forest Service was a model of centralised forest management. Similar bureaucratic structures were established in several countries, both in Asia and in other regions. This bureaucratisation of forest management, which further intensifed after independence, had in most cases an adverse impact on the livelihoods of the indigenous peoples and farmers.

The position of local forest users *vis-à-vis* the Forest Department is central to each of the three Indian case studies. The minimal allocation for forest management in India's federal and state budgets reflects the authorities' perception of forests as primarily a source of revenue and raw material for industry. Current policies such as India's Forest Law do no justice to the importance of forests in the lives and values of local people (Savyasaachi 1998).

In as far as local tree and forest management have received government attention, this has been mainly in the form of large, ambitious schemes, notably under the pretext of 'social forestry'. A clear example of this is in Karnataka, where social forestry – and, later, joint forest management projects – were adopted by the state government with substantial foreign aid financing.

People, forest and politics in Bastar

The Indian botanist Madhu Ramnath, with the assistance of *adivasi* villagers in Bastar, in the state of Madhya Pradesh, offers insight into the practical and psychological attitudes of the *adivasi* towards forests and land in Central India. The study (Chapter 1) deals both with a conflict of interests and with a clash of perceptions of forest. What is perceived by *adivasi* as 'forest' is seen by outsiders – forestry officials, lawyers, merchants and the Madhya Pradesh Communist Party – as 'land': an asset that can generate money. Ramnath argues that if outsiders start to understand and appreciate the *adivasi's* profound knowledge of the forest, there is room for reconciliation. In the course of his research, the *adivasi* have been encouraged to undertake concrete activities to protect the forest, such as fire prevention, tree nurseries, and putting a halt to the use of poison for fishing. These positive steps have helped to create common ground where the *adivasi* and the Forest Department can work together.

Chakrashila National Park

The study by Nature's Beckon. an environmental group in Assam, north-eastern India, tells the remarkable story of a group of local nature lovers who joined forces with seven *adivasi* communities to oppose the destruction of the Chakrashila Hill Reserve by illegal logging and commercial poaching (Chapter 2). It describes 16 years of diligent and courageous work which eventually led to the current situation where an NGO and local people are co-managing a national park. This arrangement is unheard of elsewhere in India. Perhaps the most promising feature of this success story is that the *adivasi* communities living adjacent to the forest reserve have been able to reduce their own use of the forest for cattle grazing, hunting and fuelwood, while at the same time improving their livelihoods by developing agricultural techniques and through the sale of handicrafts.

The Halikar Village Forest Committee

In the Halikar case study (Chapter 3), the NGO Prakruthi acknowledges that there is some value in the objectives of the 'joint forest management' concept. But the authors argue that the implementation of such schemes in Uttara Kannada District of Karnataka in the Western Ghats mountain range tends to be in conflict with the interests of local people and may even damage the forest ecology. Instead of serving basic needs and strengthening local capacity to use and preserve forest, control is taken away from local people, and their preferences and needs are disregarded. In Karnataka, where the State Forest Department is implementing joint forest management sponsored by British Overseas Development Assistance (ODA), funds are being used for commercial monocultural tree plantations to produce cheap raw material for the paper and rayon industries. In the process, the department has even displaced landless farming families who allegedly have 'encroached' on forest land.

Prakruthi describes the vigilance of people in Halikar Village who, notwithstanding these political and economic pressures, have managed to maintain control over their communal forest since 1924. But the Halikar Village Forest Committee is at a crossroads, facing increasing individualisation and economic pressures that draw people away from the village and the forest. Prakruthi argues that it is vital for the Committee, as with so many local institutions elsewhere, to find a response to these changes in order to ensure that local people continue to identify with and take care of their forest.

People's forests in the Cordillera

The next case study (Chapter 4) is by the Montanosa Research and Development Centre in the Philippines. It describes the extraordinary resilience of local indigenous communities that have witnessed the destruction of their forests by mining and commercial logging on the island of Mindanao. As with the Assam case, this study may convince those who still maintain that local people destroy forests that, on the contrary, local people are often the only players one can count on to protect these precious ecosystems. One of the communities with whom the Centre collaborates actually reforested an immense watershed that had been denuded by heavy commercial logging during the Marcos years. This study portrays the long battle of the indigenous peoples living in the Cordillera to protect their land against further exploitation by outsiders and points a way towards the consolidation and strengthening of local forest management.

South America

Centro El Encanto

In Esmaraldas province, Ecuador, Acción Ecológica collaborates with Chachi communities to examine alternative methods of forest use for subsistence and income generation (Chapter 5). The NGO and the communities are buying time. Pressures from logging companies that want access to Chachi forest lands are mounting, creating confusion and conflicts among communities as some groups succumb to offers of immediate cash for timber. The case provides a full overview of the many interests at stake, and elaborates the differences in position and attitude between local men and women. It hands down a list with concrete options and tasks to be taken up.

Yvytyrusu Forest

During the military dictatorship of General Stroessner and the years following his deposition, Paraguay lost more than half of its forest. Deforestation reached 600,000 hectares per year. As forests were laid waste and land was appropriated, thousands of indigenous people were murdered and some peoples became extinct. Internationally, these tragic events have somehow escaped notice. Yet the persecution continues, with the recent destruction of the ancient forests of the Mbawi and Ache peoples by colonists and cattle rangers. In this case study, Sobrevivencia, a Paraguayan NGO, describes its attempts to support local forest management in this difficult context (Chapter 6). It proposes concrete policy steps, based on systematic research, to reclassify and protect substantial areas of forest and to acknowledge specific ancestral land rights.

Juréia-Itatins Ecological Station

The fragility of most forest ecosystems is matched by the vulnerability of the accumulated knowledge about forest ecology possessed by forest-dependent cultures. Inspired by the adage 'We will not preserve what we do not know', the Brazilian NGO Vitae Civilis embarked on a long-term collaboration with local communities in the Mata Atlantic rainforest in Sao Paolo state. The purpose of this collaboration is as much to preserve the Mata Atlantic rainforest – and local peoples' knowledge of its ecology and sustainable potential – as it is to promote the rights and well-being of the area's indigenous, Afro-American and white settler communities. The Vitae Civilis study documents local peoples' needs and aspirations and details the shortcomings of current legal arrangements (Chapter 7).

'Washing hands with soil' in Costa Rica

Costa Rica has often been referred to as the Switzerland of South America. The country has been spared the violence, political instability and economic decline endured by the people of neighbouring nations. It has a low population density and its rich soils are favoured by an extraordinarily mild climate. Tourism is the country's most lucrative source of income. It is said that Costa Rica's greatest wealth is embodied in the richness and beauty of its nature, and in response to this widely held belief 25 per cent of Costa Rican territory has been given protected status. Yet, in spite of these favourable conditions, the rate of deforestation in this country is 250,000 hectares per year – among the highest in the world.

In their case study (Chapter 8) the farmers' cooperative Arbofilia, with the assistance of the NGO Asociacion Ecologista Costarricense (AECO) and the University of San José, grappled with this apparent contradiction. Their case study depicts a situation in which the combination of land accumulation – notably in the hands of foreign plantation companies – deforestation and depopulation of the countryside has led to the erosion of both ecological and cultural diversity. From Arbofilia's work in the Central Western Province and the Osa Peninsula come suggestions to empower communities and to 'enlarge local peoples' ecological, economic and cultural space'.

Central-West Africa

The forests of Central-West Africa constitute the world's second-largest area of contiguous tropical forest (after the Amazon basin). While deforestation has hit some countries in the region, like Ivory Coast and Ghana, the rainforests of the Democratic Republic of Congo and southern Cameroon remain largely intact – although commercial logging and road building are currently moving into these areas.

By comparison, in the Amazonian countries a broad movement encompassing academicians, NGOs, local interest groups and indigenous peoples emerged as early as the 1960s. They began to oppose the ongoing forest destruction and human rights violations that were sweeping the Amazon basin. Such a movement is still in its infancy in Central-West Africa. But it is evolving. Three organizations – from Cameroon, the Democratic Republic of Congo and Ghana – pioneers in local forest management in their respective countries for many years, contributed case studies.

Bakas, Bantus and the forest

The Cameroon government has expressed the ambition to become Africa's leading exporter of tropical timber. At the same time, international donor agencies and conservation organisations have persuaded the government to set aside large chunks of forest as protected areas. The NGO Centre pour l'Environnement et le Développement (CED) in southern Cameroon describes how the country's rainforest is being carved up, leaving many local communities empty-handed (Chapter 9). Drawing on legal studies and experience in working with Bantu and Baka communities, CED presents a number of concrete suggestions. It argues that community forestry is well defined in the country's Forestry Law, with references to a host of local management practices. The study emphasises that it makes sense from a conservation point of view to entrust forest dwellers with distinct management and user rights over the forest.

Kahuzi-Biega National Park

The Democratic Republic of Congo contains over half of Africa's tropical moist forest. In the second African case study in this volume, the Democratic Republic of Congo's NGO Centre d'Action pour le Développement Durable et Intégré Dans les Communautés (CADIC) describes the conflicts arising between the park authorities administering Kahuzi-Biega National Park in eastern Democratic Republic of Congo and the local population (Chapter 10). They explain how park management drastically curtails communities' user rights over land and forest. CADIC also notes that traditional forest management practices fail to cope with the increase in population density and the over-exploitation of wildlife, and suggests new avenues for the reconciliation of conservation and local peoples' needs.

Meanwhile, another reality has entered the forest. The case study area, located in the province of South Kivu, has become a setting for civil war. In the forests adjacent to Kahuzi-Biega, thousands of Tutsi refugees have been trapped by the cold and lack of food and drinking water.

Boti Falls Forest

Logging, over-exploitation of forests for fuel, and conversion of forest into agricultural land continue to threaten Ghana's forests. In 1996, the Ministry of Lands and Forestry launched a forestry development master plan that defines policy direction for the coming 25 years. But the plan does not address the fundamental problems of deforestation in Ghana. While the master plan's goals and objectives seem comprehensive in scope, the core of the matter is timber, to which other forest functions such as non-timber forest products, wildlife, regulation and the protection of water and soils are subordinated. Ghana's forest management strategies, as outlined in the master plan, are not based on a country-wide land-use policy and lack clear linkages with other sectors that have competing demands on forest and forest land. These are the same shortcomings that crippled official 'master plans' and forestry blueprints in other parts of the world.[1] Furthermore, the plan downplays the problem of wastage during logging. The Timber Export Development Board recommends an allowable cut of between 600,000 and 800,000 cubic metres per year, while the master plan maintains a target of one million cubic metres. Given the fact that the sector suffers from an estimated wastage of 50 per cent, 1.5 million cubic metres of timber need to be logged annually to reach this target.[2]

For many years, NGOs in Ghana have called for a reorientation of forest policies, and have emphasised the need for the protection of remaining forests in order to safeguard the country's ecological stability, rural welfare and natural riches. A number of groups have initiated pilot projects in villages to support forest management that takes advantage of the various social, economic and ecological functions of the forest. In Boti Falls near Accra, the Ghanaian NGO Green Earth Organisation (GEO) demonstrates the benefits of involving different groups – farmers, the Department of Tourism, the Forest Department and journalists – in the preservation of forest (Chapter 11).

North America and Europe

Whereas increasing attention in the past decade has been paid to local forest management in the tropics, not much has been written about similar experiences in Northern countries. Most old-growth forests in Scandinavia and Western Europe have been destroyed. In North America, where large tracts of natural forest are still intact, governments sponsor multinational companies to log these areas. And, whereas deforestation in tropical countries is met by increasing international concern and criticism, the governments in the North seem to get away with irresponsible policies towards their own forests and forest-dependent communities.

As in other former colonies, the organisation and aims of forest management in the United States and Canada were rooted in the feudal assumption that 'the state's rights to and interests in forests hold primacy over those of local people, and that the state's objectives for forest resources need not be determined by – and may run counter to – the interests of local people' (Henderson and Krahl 1996).

Following independence, the United States federal government acquired more than 571.4 billion hectares of public land, of which over a third was forested. Following the physical removal of native peoples, land was allocated to states, corporations and individuals to facilitate rapid development. At the end of the last century, in response to public concerns about accelerating forest destruction, substantial national forest reserves were placed under the exclusive authority of the US Forest Service. Gifford Pinchot, the first Chief of the Forest Service, stated that 'National Forests are made for and owned by the people [and are to be] managed by the people' (Henderson and Krahl 1996). But the Forest Service did not create an opening for local communities, forest users and other citizens to participate in actual decision making and management.

Instead, the governments of the United States and Canada tend to give preferential treatment to large-scale industry, in terms of both subsidies and the allotment of state forest resources. Local communities are thus denied access, and forests are damaged by unscrupulous exploitation. The three North American case studies also reveal a mismanagement of taxpayers' money and the destruction of jobs and job opportunities. The studies also offer hope, however, with concrete examples of how local communities can respond creatively to these social and ecological misfortunes.

The Confederated Tribes of Warm Springs

In the US state of Oregon, the Confederated Tribes of Warm Springs inhabit a 261,500-hectare reserve in the Deschutes River basin, of which about 161,500 hectares are forested. The forested area includes 27,500 hectares of old-growth forest, now a rarity in the Pacific northwest. The tribes have practised clearfelling for the last 50 years, and timber exploitation remains an important source of income. In the 1980s the tribes started to become concerned about the declining quality of the environment, fearing that logging was having a negative impact on their future.

The study (Chapter 12) focuses on the implications of a recent decision by the tribes to adopt an integrated resource management plan that balances the optimisation of timber harvest and economic revenues with protection of biological diversity, watersheds and water resources. This is also perceived as a way of safeguarding the production of food, which is both hunted and gathered. The plan requires the tribes to reduce the allowable timber harvest by 50 per cent, thereby forgoing some US$4 million in annual revenue. They are also required to spend an additional US$1.2 million in meeting stipulated environmental and social standards.

In this case study the NGO Environmental Defense Fund (EDF) and the Confederated Tribes of Warm Springs give a detailed account of the political process by which these decisions were reached, and the social and technical choices made during planning. The case study presents an important model to inspire and instruct sustainable forest management elsewhere in North America.

Working with the woods

New Brunswick is the Canadian province with the longest history of intensive commercial logging. Half of New Brunswick's forest is government-owned land, which is licensed to eight companies, six of which are transnationals. Over the past 30 years employment in the logging and wood-processing sectors has dropped dramatically, while natural forests are being liquidated to supply the massive fibre needs of the pulp and paper industry. Clear-cutting is by far the most prominent harvesting method.

On the east coast, the Christmas Mountains – so named for peaks that resemble Santa Claus's reindeer – are traditional Micmac and Maliseet First Nations territory. The land was never ceded or seized by conquest, and represents the only old-growth forest still remaining in New Brunswick. Already in 1992 a coalition of First Nations, scientists, NGOs and fishermen, together representing 20 per cent of New Brunswick's population, pleaded for a moratorium on logging and road building in the Christmas Mountains. This appeal was ignored, and in 1992 the 20,000 hectares of unfragmented forest were licensed for logging. Today only 4,500 hectares remain.

The Micmac, Maliseet and Passamaquoddy peoples, together with citizens' groups such as Friends of the Christmas Mountains and the Conservation Council of New Brunswick (CCNB), are continuing their opposition to the destruction of what is left of the Christmas Mountains forests. In this study (Chapter 13) CCNB draws attention to three promising examples of local forest management which share the rationale that strengthening existing examples of community-based forest management is a first step in the transition from centrally planned forest use to community forestry.

The Algonquins of Barriere Lake, Quebec

This case study (Chapter 14) also records the fight of an indigenous community that is striving to obtain government recognition of their management rights over customary hunting territory. At stake is a trilateral agreement between the provincial government, the federal government and the Algonquins to collaborate on integrated management of the Algonquins' land. The case study portrays the vigilance and patience of a people – having already suffered great physical and mental trauma following the invasion of their land by logging companies and sports hunters – in the face of divide-and-rule tactics and obstructive manoeuvres by the government. The Algonquins, nevertheless, have chosen a proactive approach and continued research and consultations to achieve balanced land-use planning and forest management.

The Niepolomnice Forest

Within the European context, the situation of Poland is rather unique.

Probably more than any other country in Europe, Poland has been able to preserve substantial tracts of natural and semi-natural forest. For various reasons, the country has not yet plunged into large-scale industrial agriculture. As a result, Polish forests are part of a historical landscape that is still dominated by hundreds of thousands of smallholder farming enterprises.

But Poland is currently experiencing a period of rapid economic and political transition. Privatisation and the growing influence of the European Union and other external forces have already had a major impact on Polish agriculture. Based on a detailed study of one of Poland's largest forest complexes, the Niepolomnice Forest, the NGO Polish Ecological Club (PEC) contributes a timely and realistic analysis of the opportunities and threats that socio-economic and institutional changes present to Poland's forests (Chapter 15).

The woodlot owners' association of Stramproy

In Limburg Province in the south of the Netherlands, the Educational Training Consultants (ETC) collaborated with the Forest Owners' Association ('Bosgroep') of Limburg and Both ENDS in preparing a Dutch case study (Chapter 16). Although approximately 40 per cent of Dutch forests are privately owned, this category of forest receives little attention. The study by ETC stresses the potential role of these private forests as natural refuges in one of the most densely populated and industrialised countries of the world. For Dutch nature to survive it must be offered every possible niche. The study records ways in which collaboration among the woodlot owners and local authorities helps to improve forest management.

The Udege experience in the Russian Far East

The global consumption of wood-based products is expected to increase dramatically over the coming two decades, while the yield from existing secondary forests is fixed. The 'timber-mining era' is coming to an end in Scandinavia, eastern and central Canada and most of European Russia, as very little primary old-growth forest is left. Current targets for expansion of old-growth logging in the North are the remaining old-growth forests in western Canada and parts of Russia (Olssen 1995).

The conversion of Russia into a raw material colony for the global wood products industry proceeds. This process is accelerated by bilateral development programmes, World Bank projects, forestry planning by Western consultants, and Western and Japanese investments in joint ventures. Wasteful and ever-increasing paper and wood consumption in Western Europe, Japan and North America is one of the major driving forces (Olssen 1995). Overnight, the indigenous Udege people in the Russian Far East (eastern Siberia) have started to face an influx of Korean, Malaysian and North

American companies into their territory, the Bikin region. These firms take advantage of the political and economic vacuum in Russia to cajole local authorities to grant them million-hectare logging concessions. The authors of the case study in Chapter 17 give an account of local economic strategies – based on non-timber forest products – undertaken by the Udege people, with the back-up of sympathetic organisations, to maintain their economic autonomy and to present a viable forest management alternative to destructive logging scenarios.

NOTES

1 Based on FOE–Ghana's 'Forestry Review', in *FOE-line*, No. 5 (July–September 1996), p. 61.
2 *Ibid.*

BIBLIOGRAPHY

Henderson, D. and Krahl, L. (1996) 'Public Management of Federal Forest Land in the United States', *Unasylva*, Vol. 47, No. 184.
Olssen, R. (ed.) (1995) *The Taiga Trade. A Report on the Production, Consumption and Trade of Boreal Wood Products,* Taiga Rescue Network.
Rietbergen, S. (1993) *The Earthscan Reader in Tropical Forestry*, Earthscan Publications Ltd, London.
Savyasaachi (1998) *Tribal Forest-Dwellers and Self-Rule*, Indian Social Institute, New Delhi.

1 People, Politics and Forest Management in Bastar

An Ethnobotanical Perspective

MADHU RAMNATH

India

This case study describes the relationship between people and forests in Palob Village, the main study site, in Bastar District in the state of Madhya Pradesh in Central India. Bastar is still a densely forested, sparsely populated area, home to one of the world's largest concentrations of *adivasi* (or indigenous) populations. These people depend for their sustenance on the forest, most of which falls under the jurisdiction of the wildlife and revenue sections of the Forest Department. Regulations curtail *adivasi* livelihood and contrast sharply with the indigenous people's ways of managing forests. The *adivasi* practice of shifting cultivation also contradicts the state's views about 'rational' forest use. Unusually, other religions have intervened little in the Hindu-tribal dynamics of the region, so that comparisons can be drawn quite clearly between traditional knowledge and beliefs, on one hand, and conventional views on the other. The study, based on detailed ethnobotanical research conducted in Bastar villages, raises the following questions: (1) How is the forest space perceived by the *adivasi*? (2) How is the forest space regarded by the administration? (3) What are the areas of conflict/dialogue between the people and the authorities?

Forest, Geography and Climate

Central Bastar is in the south of Madhya Pradesh State, bordering Orissa. Tropical deciduous, *sal*, mixed and teak forests are intermingled with bamboo. The greater part of Bastar is a plateau, which varies in elevation between 284 and 1,200 metres above sea level. The average annual rainfall in Bastar District is 1,538.4 millimetres, of which about 86 per cent occurs during the monsoon months (June–September).[1] The forest in the area around Palob Village varies from moist peninsular *sal* (*Shorea robusta*) to tropical dry, one sometimes contiguous with the other. There is very little teak here, except in plantations; in moister areas, *sal* dominates and gives its peculiar character to the forest.

41

The *sal* (*sarangi, arang, sargi*) tree is both sacred and important to all the *adivasi*. Its leaves and the resin it yields, which is used for incense, are part of all religious ceremonies and rituals. Its wood, hard and durable, is used especially in building. Its seeds are gathered in summer, cooked and eaten. It is also an important minor forest product that has been 'nationalised', giving the Forest Department a monopoly over its collection and sale. In recent years *sal* seeds have been in high demand; the Department buys them from people by organising centres where they can be collected.

Where the canopy is continuous and there is little undergrowth, the *adivasi* call the forest a *malang*. This denotes a condition of the forest and not its type. *Sal malangs* are important for gathering fibres, tubers, mushrooms, leaves and twigs that people use in everyday life. When a patch of forest is cleared for cultivation and later abandoned, it goes through successive stages of vegetation before it returns to the *malang* condition. The earth that was 'tired' from being cultivated is said to have recovered when a *malang* appears again.

Where *sal* does not dominate, as on some hillslopes, the vegetation is varied. There are trees of the *Terminalia*, *Adina* and *Pterocarpum* genera, often interspersed with bamboo. In the gullies and along streams there is an abundance of mango and *Diospyros peregrina* (the fruit of this tree is relished by monkeys), along with the drooping clumps of the *pita* bamboo, *Cephalostachyum sp*. In rocky areas and on dry hilltops *sal* is usually absent. Instead, there are tree species – many with pale bark, like *Sterculia*, *Gardenia*, *Xylia*, *Phoenix* and *Euphorbia* – that are resistant to summer heat. Certain species of bamboo thrive in these dry conditions.

Though many plants are found in several different parts of the forest, some are specific to a single area. Efficient gathering therefore demands detailed knowledge of floral distribution in the forest. For instance, if one sets out to collect silkworm cocoons, it would be in an area with a lot of *mange* (*Terminalia arjuna*), *sal* (*Shorea robusta*), *nammi* (*Anogeissus latifolia*) and *merdengi* (*Terminalia tomentosa*) trees, and in that order. As local people say: 'The moths find *mangi* leaves the sweetest.'

All springs, streams and rivers are marked by rituals and sacrifices addressed to the spirits of water and other parts of the forest. Caves, unusual rock formations, giant trees, the meetings of streams and small and large waterfalls in the region are all, likewise, sacred places on the *adivasi* map.

The People and the Forest

Palob Village, where I lived during the course of this research, is surrounded by reserved forest, a national park and teak plantations, all of which are managed by the Forest Department. Almost 70 per cent of the people living in the area are fighting cases of 'encroachment', filed against them by the Department, because they have occupied forest land for cultivation. Most other villages in the region are under similar judicial circumstances, causing a steady drain on people's resources.

The emphasis on plants in this research and the establishment of a rudimentary nursery drew the interest of local people. As there were plants in the nursery from several parts of the forest, including some exotics from outside the region, both men and women would drop in to see them. Some plants were brought to the nursery by people who had come across them and thought that they could be of interest. The identities and uses of several species were ascertained during these informal chats.

Plants pervade most conversations in a forest village. The flowering and fruiting of plants, the appearance of mushrooms in the *malangs*, the shooting of new bamboo and tubers that are ready to be dug up are all matters of immediate concern. Each of these events is followed in detail by everyone, and their larger implications discussed.

The term *chularana*, meaning 'to wander', describes the daily journey of an *adivasi* in the forest. There may be specific things to be gathered, like leaves or bamboo shoots, or crabs to be caught, but the exact route of the journey unfolds while on the move. In a village like Palob, which has sufficient forests, people wander within a radius of five kilometres for their daily requirements. This can increase to 15 kilometres during hunting and fishing expeditions, or when searching for particular plants. Still longer journeys are made to visit relatives or markets, and to participate in hunts in other forests, when invited.

Common *adivasi* uses of the forest include:

- Food, in the form of yams (*Dioscorea sp.*) and other tubers, mushrooms, bamboo shoots, greens (*Bauhinia purpurea, Amaranthus sp.*), several kinds of fruit and nuts, honey, red ants, eggs and meat;
- Fibre, from the barks of trees (*Kydia calyciana, Butea superba*) and climbers (*Bauhinia vahlii*), with which to fashion different kinds of rope;
- Medicine, for common ailments such as coughs, fevers and bruises;
- Oils, from the seeds of *Pongamia pinnata, Madhuca indica, Schleichera oleosa*; the latter two are also used in cooking and the first in medicine;
- Wood, for building and agricultural implements, musical instruments and fuel;
- Bamboo, for fences and building, and to fashion articles like various baskets, bows, arrows and flutes;
- Leaves – of *Shorea robusta, Bauhinia vahlii* and *Holarrhena antidysenterica* – from which containers are fashioned;
- Twigs, used as toothbrushes (*Shorea robusta, Pongamia pinnata, Syzigium cumini*).

Chularana also includes the casual hunts, which are sporadic and without ceremonies. These are journeys of two or more people, often with dogs, to search for small game, such as civets, flying squirrels, monitor lizards, forest rats, hares and snakes. The movement and nesting of these and other animals are also followed during these wanderings. Occasionally, women accompany men on these journeys. Women do not usually hunt meat during *chularana*, however; they are more often occupied with the gathering of plants.

Hunting, Fishing and Agriculture

Each section of the forest is considered the sacred jurisdiction of particular spirits and deities. The annual hunts, the *kedhkul*, are religious ceremonies as much as they are ventures for meat. They take place in the summer over five to six weeks, preceded by rituals and sacrifices for the resident deities. It is considered necessary to 'wake up the hunt' each year and every able-bodied man and boy is expected to participate (men absenting themselves from the *kedhkul* are liable to be smeared with mud by the women, and fined by the women's council). Game includes wild boar, porcupine, bear and deer. The animals brought back from these hunts are neither skinned nor disfigured in any way, 'as it spoils future hunts and brings ruin to the village'. Only during the *kedhkul* do people go to all the different parts of the forest in their particular territory. On these trips, the rites due to each sacred spot are performed in turn, and young people begin to learn about the distribution of trees, the movement of animals, water sources and exceptional places during these expeditions.

A favourite activity is fishing. Though nets, traps and rods are used by some people, the more common method practised is damming small streams and bailing water through a meshed basket. Plant poisons (*Derris sp.*, *Randia sp.*) are also used where the water is neither swift nor deep. These are usually group affairs, with children often joining in. Entire days are spent along streams in summer, catching fish and crab and gathering edible ferns and other greens (*Colocasia esculenta*, *Ipomoea aquatica*).

There are five kinds of areas used for cultivation among the *adivasi*: *wada* garden patches within the fence; *wayal* fields that retain some water even during the dry season, though usually not enough to grow a crop; *jodkil* fields that do not retain any water at all during the dry season, and are usually less fertile than the *wayal*; *manom* cultivable patches made by clearing the forest; and *penda* cultivable patches made by clearing the forest on hillslopes.

Almost everyone has a *wada*, but very few people have a *wayal*. One could almost deduce that owners of a *wayal* have longer ancestry in the village, as they often also own the larger of the tamarind trees in the village, planted by their parents and grandparents. The sale of tamarind brings in considerable income each year.

Cultivation takes place mainly during the monsoon, as most of the rest of the year is too dry for agriculture. Within the *wada*, vegetables like cucurbits, chillies, tomatoes, bananas and papayas are grown. Some grow plants valuable for their roots, like turmeric, ginger, *beska* (*Costus speciosus*), colocasia (*Coloasia esculenta*) and some species of yams (*Dioscorea sp.*). Those who have larger *wadas* grow maize (*Zea mays*) and a cash crop of mustard/sesame at the tail end of monsoon. The *wayal* and the *jodki* are used mainly to cultivate rice (*Oryza sativa*) and sorghum (*Panicum sp.*).

The traditional practice, before new *manoms* are cleared, was for the men of the village to sit together and discuss the site. Sometimes a part of the village, or the whole village, would move closer to the new *manoms*, but this was less likely with a large village. The *manoms* themselves are long, narrow strips of cleared forest, preferably near streams from which water can be channelled for use during the monsoon. Almost as a rule, two *manoms* are separated by forest. When 'the earth is tired' – also expressed as 'the earth is menstruating' by *adivasi* in west Bastar – after three to six years, the *manom* is abandoned and a new one cleared. An additional motive to move from an old *manom* is the increase in rats and other pests that destroy the crop. The unused *manom* is quickly taken over by secondary forest.

The *manom* and *penda* are most fertile in their first years and are used to grow rice, millets, and a variety of other edible plants. Saplings of chillies and tomatoes are raised in the *manom*, before being transplanted elsewhere. The *manom* is in many ways an 'experimental space' within and surrounded by the forest. When the crop is green it has to be guarded against herbivores at night, and *machans*, or platforms that serve as lookout posts, are erected to spend nights in the *manom* itself. When the seeds are ripe the crop needs protection from birds during the daytime. After a few years, when the fertility of the *manom* is exhausted, it is abandoned and allowed to recover, and a new *manom* is sought.

All cultivation, whether in the village or in the forest, is preceded and guided by rituals. The season terminates with the harvest brought from the threshing floor to the home, coinciding with the festival of *Dilwel*, when drums mark the passing of the season.

Customs and Tensions

Many of the laws of *adivasi* tradition pertain to their relation with the forest. Laws relating to the harvest of uncultivated plants refer directly to their life cycle. When a leaf may be used to make leafcups, when a fruit may be plucked or eaten, and details about when and how various foodstuffs may be cooked are all part of this tradition.

Meddul, for example, the Festival of Mangos, is a ceremony that directly relates to propagation of the mango (see Box). As the *adivasi*, by and large, do not cultivate what they find in the wild, these rules are important for regeneration. The festival is quite different in villages near towns and in places where the Hindu influence is felt. Here the dates for the festival (and for the harvest of mangos) are set by Hindu priests, who consult an almanac and not the mango tree, often with an effort to combine the *adivasi* customs with a Hindu festival. Mango harvest dates in these towns may precede those in villages by up to three weeks. This trend is also influenced by traders who buy raw mangos to supply the pickle industries.

Mangos

When the first mangos begin to ripen on the trees talk of *Meddul*, the Festival of Mangos, fills the village. Wind-dropped mangos lie under the trees, but even children do not show an inclination to pick them up. The priest of the village, along with other elders, observes the fruiting and ripening of mangos in the forest and fixes the day 'to taste the mangos'. By then the monkeys and bears and many birds have already tasted the fruit.

The actual dates for the festival may vary from village to village, but fall within the same period. A criterion that seems to be taken into account is the availability of mature seeds for regeneration. A good date would allow the maximum possible fruit to ripen on the trees, while at the same time ensuring that animals do not cut too deeply into the community's share. After mango eating has begun there are still cooking rules to be observed. Mangos may not be cooked until a week later, when *Mati Tiyar*, the Festival for the Earth, is performed. Hence raw mangos (used in cooking) are allowed to ripen for another week, furthering the possibilities of regeneration.

For weeks following the festival, mangos occupy people fully. Whole days are spent gathering and eating mangos, visiting trees whose fruit are known for certain qualities. Seeds from the sweet fruit of trees that are very old are planted and cared for, and some people experiment with grafting. The hunters in the forest take long breaks under mango trees. Friends and relatives visit each other to taste mangos from other areas, and news about famous mango trees of the region spreads.

In recent years the growing pickle industry has increased the demand for raw mangos. In villages devoid of forest and nearby towns, mangos are never allowed to ripen. The trade in mango seeds – the kernel is reportedly used in pharmacy – is also a serious threat to this forest fruit. The *adivasi* from many villages gather mango seeds and bring them to the market to private traders. When asked why they participate in this seed trade, knowing as they do the consequences it could have, their answer reflects a prevalent mood: 'Whatever they put a price on will sell. If we don't take it to the market, someone else will.'

The result of these commercial forces is that town *adivasi* travel in summer to nearby forests to gather raw mangos. But the *adivasi* who live in those forests are bound by their own laws not to harvest the fruit until the festival is over. These conflicting interests within the *adivasi* communities cause violent and angry scenes in the forest.

Some of the other fruits which, like the mango, are found in the wild and not planted, have such taboos to protect them from being exploited without foresight. To break these taboos, it was said, 'would bring ruin to the village'. Children are told that 'panthers and tigers will enter the village'. Anyone found disobeying these basic tenets is liable to be penalised by the council of elders. The breaking of taboos seems not to occur among individuals in a village. It usually involves a substantial section of a village, when the elders themselves have lost control over their space and their people.

Rights and Ownership

Adivasi who have lost their forests – 'the forest has run away', as they put it – have no option but to exploit other forests. People come from up to 20 kilometres away, and on foot, to the forests around Palob. During the summer months, and for brief spells throughout the year, groups of people come to camp in these forests. They gather leaves, silkworm cocoons, and *sal* resin, and make rope, chop bamboo, fish and hunt for a few days. But as these forests 'belong' to other *adivasi*, the strangers are wary of their position. In some areas the 'owners' of the forest are paid a kind of tribute, in cash and kind, by other users of the forests.

Forest 'ownership' relates directly to hunting rights. Those with rights in a specific part of the forest are the ones who traditionally have performed all the religious ceremonies due to that space. During the annual *kedhkul* hunts, it is their privilege to 'lead' the hunt, even though they may not frequent that part of the forest. (There are instances when the 'owners' come to their forest only during the *kedhkul*, as it is too far away from the village. Daily users have no 'legitimate' rights.) In this sense, the *adivasi* people recognise ownership and tribute within their forest.

Adivasi who have to gather in forests far from their homes must also deal with the Forest Department field staff, who are not as lenient to them as they may be with people from nearby villages. *Adivasi* are rather tense while moving about in an alien forest, where they have neither traditional nor administrative rights. Little care is given to what they harvest and their aim is to carry away as much as they can. Harvest of unripe forest produce like *sal* resin gathered almost a month before its time and over-gathering of bamboo shoots and *sal* seeds are examples of careless forest use. Experiencing this treatment of their territory, within which they have no legal rights, encourages traditional owners to adopt new, more exploitative approaches.

The Market and the Forest

The main forest products are classified by the Forest Department as 'nationalised' and 'non-nationalised'. In spite of these rules, a trade flourishes in many forest products, regardless of their status. The demand for the bark of the *menda* tree (*Litsea sp.*, reportedly used for the base of incense sticks) has in recent years led to a dearth of young and mature trees in Bastar. Supplies now come from the forests in Orissa. Similarly, a few years ago the bark of the *ahl* tree (*Morinda sp.*) and the seeds of *kappa tunda* (*Entada sp.*) were in demand, and many areas in the forest are marked by an absence of these species. Fruit trees such as *Emblica officinalis*, *Buchanania lanzan*, *Mangifera indica* and *Tamarindus indica* are also affected – to give some examples of the wider problem.

Similarly, there is a trade in many birds that are captured and exported out of the region. The hill mynah (*Gracula religiosa*) has diminished in numbers. The rose-ringed parakeet and the alexandrine (*Psittacula krameri* and *Psittacula eupatria*) are both cage birds, sold in weekly markets. The racket-tailed drongo (*Dicrurus paradiseus*) and the black drongo (*Dicrurus adsimilis*) are sought for their melodious calls, while the common grey hornbill (*Tockus birostris*) has disappeared completely from the region. Animal skins from panther, tiger and deer are also part of this economy, but information on these is less readily available.

Nationalised forest products like timber, fuelwood, *sal* seeds and the leaves of *Diospyros sp.* (*tendu patta*), used in making *bidi* cigarettes, bring in a major part of the revenue and are the department's monopoly. For *tendu patta*, licences are issued to private traders. India's immense *bidi* industry is said to be 'as important to the Department as elections are for the nation'. Records show that over the last 15 years revenues from both these minor products have increased steadily. It is also noticeable that *sal malangs* in many parts of Bastar show no sign of regeneration. Plantations of *sal* have been tried by the Forest Department itself, at great expense and without much success. *Sal* resists regeneration out of the wild. With *tendu*, on the other hand, there seems to be no immediate danger, as the leaves are collected and the tree is a good coppice supplier.

Market forces trample over the *adivasi* notions of space and 'ownership' in the forest, ignoring the details of a product, caring only for quantity and revenue. Bamboo, source of a pulp used in paper industries, provides one such example. The *adivasi* recognise seven kinds of bamboo in their forest. Each of these has qualities that make them suitable for specific purposes. To build a fence one would use *pita vedri* (*Cephalostachyum perigracile*), for its small hollow and suppleness. For a basket or rainhat one would look for *kon vedri* (*Dendrocalamus strictus*), as it is most easily stripped. *Nuli vedri* is popular for flutes because of its long internodal distance, and *kata vedri* is the best for a strong man's bow. All bamboo shoots are eaten, though some are preferred to others for their taste and texture, or because there are customary methods of cooking them. As their bamboo needs are specific, *adivasi* are very precise about distribution in the forest.

In sharp contrast, the Forest Department's approach to bamboo is based purely on quantity. Bamboo is harvested in large quantities every year, with the help of *adivasi* labour. The *adivasi* notice that the Forest Department is interested in the total amount of bamboo, rather than in individual amounts of the various species. The proportions of these species, with respect to each other, are disturbed in the process.[2] The species of bamboo and their local uses are not the Department's concern.

The Forest Department

Formerly, the state forest was divided into *zamindaries*, landholdings, each managed by a *zamindar* or landowner, whose main concern was financial gain. Between 1896 and 1928 the state suffered a deficit of funds which led to over-exploitation of the forests. In 1957 the management of the *zamindari* forests was taken over by the Forest Department. But even as early as 1904, forests were classified as reserved and protected, and villages within these areas were served notice to leave the place and settle outside. More efforts were made four years later, to restrict shifting cultivation within these areas. An *adivasi* rebellion took place in 1910, in which some state officials were killed.

Today's Ministry of Environment and Forests has a Forest Division, headed by an Inspector General of Forests. All projects funded by the central government are monitored by this division. It is responsible for giving clearance for forest conservation projects, and has offices in six regions. The central government funds special projects on social forestry, joint forest management and tiger conservation. Some of these projects are in protected areas, but their implementation is the responsibility of the state governments.

The State Forest Department derives most of its funds from the state government. Senior staff are recruited from the Indian Forest Service, in any part of the country, while junior cadres come from the State Forest Service within the state. In central Bastar, forest management generally follows the Forest Department's ten-year plan.

Local People and the Forest Department

There are a few areas where the people and the Forest Department interact. Every year, during the dry season when there is little agricultural work, the Forest Department employs *adivasi* from surrounding villages to work on construction sites, road laying and the marking, felling and extraction of timber. More thought could be given to whether this paid work is beneficial in any other way to the *adivasi*. They value the daily wages, but often express the opinion that the roads are of no use to them.

Forest Department field staff – the range officer and his subordinates, the *nakadars* (forest guards) and the *chowkidars* (forest watchers) – have direct contact with local people. *Adivasi* often suffer from corruption among these officials, especially those coming from outside the region. Senior officials, though aware of the problems, are unable to check them.

The appointment of local people as Forest Department field staff is a welcome trend, as they speak local languages and are aware of local needs and customs. This makes them more lenient towards the *adivasi* who use the forest for their daily requirements. On the other hand, due to close relationships and kinship bonds in the area, they are often unable to prevent more serious crimes that may occur, like the poisoning of animals or the smuggling of forest products. These officials are caught in a dilemma: they must choose between risking their jobs or facing social boycott and possible violence. Though it may be marginally preferable to have local staff in the Forest Department instead of importing them from other states, one must keep in mind their vulnerable position in the field.

Conflicts

There are several causes of conflict over forest resources in the central Bastar region, the principal one being confusion over the rights of local people versus those of the state. The traditional use of the forest by the *adivasi* has been curtailed by law. *Nistar* rights over 'forest produce required for *bona fide* domestic or agricultural purposes', as defined by the Forest Department, were abolished in the state in 1976. Instead, *nistar* and commercial depots have been established in various places. Each depot is allotted several villages, and the inhabitants of these are expected to buy their requirements at a 'nominal' price. Some concessions include 'free collection of fuelwood from dead, dry, fallen or damaged trees, free collection of grasses, thorns, brushwood, bark (except that of *kahua*), flowers, fruit, roots, etc.' There are also concessions relating to the grazing of cattle without charge or at concessionary rates.

Fishing is a dry season activity that brings the conflict to the fore, as it is legally an offence within reserved forest. People (usually women) may go on two or three 'fishing ventures' a month. The common and traditional method of dyking streams, bailing or using plant poisons, is noisy and involves a lot of people. Plant poisons affect only the gills of certain fish and are not harmful to other aquatic life. However, these methods of fishing attract the attention of any Forest Department official in the area.

Perhaps for this reason, over the last four years chemical poisons have replaced plant poisons. Pesticides such as Gamaxyne were introduced by town *adivasi* to those living in the forests. The chemical is poured into the water in the evening, and the dead fish are collected floating downstream in the morning. Though fish tend to be rotten inside and smelly, people have half-heartedly begun to follow this silent technique. Within the span of a few years they have also noticed a depletion in variety and amount, but no move has been made by the Department or the people to counter this trend.

Hunting is another contentious issue. In June 1995 (during the hunting season) a barking deer strayed into Palob Village, and was chased and killed by the dogs and boys. Forest Department staff, who were present, immediately confiscated the deer and filed a case against two boys, who were eventually sentenced to jail. A lawyer was hired at a high price to get the boys released and there was much tension and worry in the village. As such cases can drag on for many years, people asked if I could intervene. Discussions were held with the Forest Department to put forward the people's point of view about hunting. A statement was made in Durva (the local language) and its translation presented to Department officials. The fact that the Forest Department agreed not to attend the hearing of the case is an indication of its consideration for the *adivasi* view of hunting. But the scale at which traditional methods of fishing and hunting are practised needs to be studied in greater detail before coming to definite conclusions.

In addition to differing notions of rights, fishing and hunting, forest fires can also lead to conflict over forest resources. Every year in the summer the forests of Bastar are prone to fires, which are usually started by people. 'To have grasses sprout soon after the first rain' is the most common explanation elicited. More often fires result from carelessness while people are gathering *mahua* flowers, when the leaf litter below the trees is burnt in order to make collection easier. Men also set fires to reduce undergrowth and dry litter, allowing them to walk more silently in the forest.

A positive development is that recently, for the first time in 15 years, the forests in the region were unharmed by annual summer fires. The flowering of a species of bamboo made the task of fire prevention urgent. Through several meetings in surrounding villages the people's help was sought. The success of these combined efforts by the forest people and the Department can be assessed by the excellent regeneration of bamboo from seed.

Politics and the Forest

Central Bastar covers 5,888 square kilometres, 27 per cent of which is forested. Two thirds of this is reserved forest managed by the Forest Department. In these protected areas the *adivasi* have few rights, except for the most basic *nistar* rights. More than half the people in the area are involved in legal battles, mainly due to cases of forest offences filed against them by the Department. The expenses incurred due to lawyers, regular journeys to the court and constant anxiety, constitute a steady drain on people's resources. Money for these ventures, which may last several years, comes essentially from exploiting the forest. This puts an additional pressure on the forest and contributes to the over-exploitation of all marketable produce.

In spite of the constant threat of legal entanglements, most *adivasi* have cleared *manoms* for cultivation. This 'illegal' activity lies at the core of the forest issue for all parties in the region. Politicians use the fact of people's landlessness in framing their own strategies; the Department claims that forest destruction makes its revenues suffer; for lawyers it has become an additional source of income; and village power wielders become inter-mediaries among the people, local politicians and the lawyers.

Yielding to significant political pressure in 1988, the state government legalised a portion of land that had been cultivated before 1976. This was done with the aim of preventing further 'encroachments' by the *adivasi*, especially in reserve forests. At the village level, this government decision was understood to mean that any land cleared will be legalised at some future date. More forests have been cleared since 1988, and in 1995 the government legalised lands that were occupied for cultivation before 1980. According to the Forest Department records, 589 people made an applica-tion from the Darba Range of central Bastar, for a total of 3,431 acres. Only 30 applicants received legal status for their land since the rest of the lands were cleared and cultivated after 1980.

Over the last two decades *adivasi* from the south of the region have been moving north into central Bastar, east into Orissa and west into Abujmarh. They had been displaced first, perhaps, by the iron ore mines in Bailadila, which caused pollution of many tributaries of the Indravati River.[3] Several new villages have appeared in the forests of Bastar and the adjoining state of Orissa. These villages have a high level of political awareness and the inhabi-tants are often bound to trespass into the hunting or gathering grounds of *adivasi* already living there. There is both tension and compromise in these relationships, but the pressure on the forests is alarming.

The *adivasi* face possible additional expenses if a legal case incurs. The poorest *adivasi*, who have nothing but a small *wada*, cannot afford to clear a *manom*. Those who can afford to fight such cases are from large households that are capable of gathering on a larger scale. The surplus income from the sale of forest products is then spent on court cases.

People who have no land to cultivate procure almost all their sustenance from the forest: directly by gathering, fishing and hunting, or indirectly by selling forest products like rope made from plant fibres, mushrooms, fruit, mats and baskets. Even for those who spend more time cultivating a crop, the forest provides daily essentials. It is noticeable, however, at least in villages involved in court cases (and other matters that require bribes), that people are forced to sell increasing amounts of forest products.

Conclusion

At this stage of the research, most of the factors that contribute and relate to the sustainability of the region have become clear. Market forces, official laws and politics directly influence the conservation of Bastar's forests. These, in turn, affect the *adivasi*, whose freedom and rights to use the forest have been curtailed. State conservation laws do not take into account the *adivasi* way of life, which depends on the forest. This leads to conflicts between the people and the administration.

Many villages that are surrounded by forests, like Palob, have experienced changes during this generation. Since the 1960s teak plantations have replaced *sal* forests in some places. Reserved forests border most 'revenue villages' (those that pay a tax on agricultural land, as opposed to those designated as 'forest villages'). In 1982 a 200-square-kilometre national park was declared, further restricting *adivasi* rights. 'We are thieves in our own forest now' is an *adivasi* opinion that expresses their status within a protected forest. There have been no attempts on the part of the Forest Department to explain the restrictions to people, or to understand how these affect the *adivasi*.

According to *adivasi* law, only during cultivation are *manoms* considered to be private land. When they are abandoned they return to the public domain. By its very nature a *manom* belongs to the forest and is not permanently owned by anyone. There are many gathering rules that confirm this. Mushrooms, bamboo shoots, tubers and greens that grow wild may be gathered from a *manom* by anyone. The cultivator has no exclusive rights to uncultivated plants even on his own land. Also, the clearing of trees for a *manom* does not give an *adivasi* any rights over the fallen trees. It could be anyone's wood. The *manom* is different from 'land', as understood by local politicians and the government. 'Land' pertains to something permanent and exclusive, regardless of what one does with it. A *manom*, on the other hand, cannot be owned permanently or completely, even while it is being cultivated.

In the conflict between people and the Forest Department in central Bastar, the popular political party in the region, the Communist Party of India, claims to support the people. The party addresses the 'land for the landless' issue and claims to fight for people's legal rights over forests they have cleared and occupied. The politicians' idea of land clearly refers to sedentary agriculture, and so for them the answer lies in securing for the *adivasi* permanent rights over their *manoms*, where they would settle down.

This apparent misunderstanding about what a *manom* is causes it to be equated to 'land'. Thus for politicians, 'land', not 'forest', is the central issue. But the *adivasi* do not see their conflict with the authorities as one over 'land'. For them it is a forest conflict. It is the difference between 'land' as a two-dimensional field and the three-dimensional 'forest' of the *adivasi*.

Meanwhile, the Forest Department does not recognise people's rights in any part of the forest. This official attempt to protect the forest denies local people's traditional rights. It is important to remember, however, that traditional rights are 'restrictive rights'. For example, though one may have rights over the mangos in a certain forest, one has to wait for the festival before plucking them. An element of conservation (or restriction) is inherent in the way a plant or its product may be harvested and used.

By failing to see this aspect of forest use in *adivasi* culture, the state effectively loses potential forest guardians, making the Forest Department the sole agent of forest conservation. This is of course impractical, and leads to the abuse of forest resources. By doing away with the traditional guardians, the forest actually becomes open to all, regardless of spaces that 'belong' to certain *adivasi*, who lead the *kedhkul* and perform the sacred ceremonies. As the people of Palob see it, even if they do not want to participate in the sale of some forest produce, they cannot stop other people from doing so. They too join in the trade, though they are fully aware of the consequences. For their part, they do not distinguish between the status (nationalised or non-nationalised) of the product. Nor do they care whether the buyer is the Forest Department or an individual, or whether the trade is legitimate or clandestine.

There is no easy way to assess the effects of the money economy on the *adivasi*. As many of the payments they make, in cash or kind, are in the form of bribes, the information is not open. Rough calculations made for a few households fighting court cases for land suggest, however, that the expenses involved can be almost as much as a family's land produces. The time, energy and tension involved in dealing with non-*adivasi* bureaucracy and law cannot be overstated.

The *manoms* cleared within the last ten years around Palob have been unusually large (up to five hectares), without any vegetation left standing. Such cultivated patches resemble the areas clearfelled by the Forest Department for their plantations. These areas were cleared collectively in the surrounding reserved forest and national park, and are within what Palob considers *its* forest. When they are abandoned the surrounding forest cannot recover completely. The inner portions of such tracts have become thickets of *Lantana sp* and *Eupatorium sp* within a few years, smothering the saplings of indigenous species. These also catch fire easily during the dry season.

The manner of forest exploitation in most parts of Bastar is no longer guided by traditional rules and customs. Trends in fishing, hunting, gathering and the clearing of new *manoms* show an extremely short-sighted attitude, and are most prevalent in regions where traditional rights have dwindled. State conservation policies that do not recognise traditional *adivasi* ownership of the forest are at the root of the conflict between people and the Forest Department. The situation then becomes political. The *adivasi* notions of forest rights and 'ownership' are neither exclusive nor permanent. But the erosion of traditional restrictions allows the *adivasi* to exploit any forest produce that is marketable, without regard for the regenerative capacity of the product.

In the interior villages, where the Forest Department has not undermined the traditional *adivasi* methods of forest use, these traditions continue to be respected. This traditional knowledge about plants and different types of forest in the region has much to offer to our understanding of conservation in tropical forests. It is likely that we will have to review conservation laws in these areas, and recognise traditional forest rights. The Forest Department's monopoly over forests alienates people from their living space, and can only hasten their destruction.

Recommendations

Forest fire prevention in the summer months should be viewed in a long-term perspective. This means seeking the support of local people, and finding new ways to communicate and combat the problem. This could eventually be expanded to a larger area with huge benefits for soil and vegetation. A more immediate task is to protect the bamboo saplings that regenerated in recent years from new forest fires. The Forest Department is prepared to support any such efforts.

Rivers and streams must be protected from chemical poisons used both by locals and by people from outside the region. This is an annual summer activity, each year leading to further reduction in the quantity and variety of fish. Serious efforts need to be made to reach out to river users for their help in stopping the use of chemicals. The possibility of concrete agreements between the people and the Forest Department regarding the use of rivers should also be investigated.

There are often abandoned fields around *adivasi* villages where little or no cultivation takes place. These areas are full of weeds and thickets that smother useful forest species, and easily catch fire in the dry season, destroying the ground litter. With the people's help, these fields could be identified. Some of the plots could be used to plant indigenous species of plants and the fruit trees of people's choice. This would mean collecting seeds and saplings, clearing the fields of weeds, and watering and protecting the plot for a few years. A scheme of this kind would provide income for local people and, if carried out successfully, would boost forest regeneration.

A broader and deeper knowledge of the role of plants in *adivasi* life is necessary to assess the full meaning of 'forest products' in *adivasi* terms. Rituals and customs related to plants, as well as notions of 'ownership' and *manom*, also need to be understood more profoundly. A study of these issues in other villages would be useful for purposes of comparison. One could then consider ways in which the state could hand over forest 'ownership' to the *adivasi* people.

A larger study is needed of plants and customs that guide the time and manner in which they are used in *adivasi* life. It has been mentioned already that many of these customs have an ecological bearing, and may contribute to the survival of these plants in the wild. The forces that undermine *adivasi* traditions need to be clearly recognised, and their detrimental effects on the forest addressed. Botanical studies could be conducted of other regions in the district to work out the minimum area needed to maintain maximum floral diversity. Assessments of land use and sustainability could then be made for a larger area.

NGOs and interested individuals should set up a legal centre to help *adivasi* with legal advice. This would enable *adivasi* to deal with their legal problems without unnecessary tension and expenditure. Such an effort would speed up procedures and bring the judicial dead-ends faced by the *adivasi* into the open.

NOTES

1 In the summer months, before the monsoon, there is rain in the form of thunder showers, usually in the afternoons. The average number of rainy days (more than 2.5 mm rainfall) in the year is 75, with little variation from year to year. From the central region of the district, the rainfall increases towards the northwest and southeast, and decreases towards the northeast and southwest. The forests in the district are more pronounced in the maximum rainfall zone. (Figures based on the records in the India Meteorological Department, New Delhi.)

2 An excerpt from a work plan reads: 'In the best quality bamboo area the crop is dense and clumps are generally congested. Many of the clumps have more than 100 culms. Clumps with culm number ranging from 40 to 70 are generally found in all such areas. In such clumps the percentage of dry and rotten culms is more, obviously due to non-working or partial-working in these areas. In the best areas the average culm height varies from 10 to 15 metres.'

3 These mines are part of an Indo-Japanese joint venture.

BIBLIOGRAPHY

Ali, Salim (1979) *The Book of Indian Birds,* Bombay Natural History Society and Oxford University Press.

Central Bastar Forest Division (1988) *Working Plan for Central Bastar Forest Division and Tongpal Range of South Bastar Division*, vols 1–2, Jagdalpur Circle, Bastar District, Madhya Pradesh. For the Period 1988–89 to 1997–98, volumes 1 & 2.

Central Bastar Forest Division (1988) *Supplement to the Second Preliminary Working Plan Report for Central Bastar Forest Division and Tongpal Range of South Bastar Division,* Jagdalpur Circle, Madhya Pradesh.

CSIR (1948–76) *The Wealth of India: a Dictionary of Indian Raw Materials and Their Products,* 11 vols, Section 1, Raw Materials, CSIR, New Delhi.

Dastur, J. F. (1983) *Medicinal Plants of India and Pakistan*, D. B. Taraporewala Sons & Co., Bombay.

— (1983) *Useful Plants of India and Pakistan*, D. B. Taraporewala Sons & Co., Bombay.

Goethe, J. W. (1978) *The Morphology of Plants*, Wyoming.

Haines, H. H. (1916) *Descriptive List of Trees, Shrubs and Economic Herbs of the Southern Circle, Central Provinces*, Pioneer Press, Allahabad.

Janzen, D. H. (1975) *Ecology of Plants in the Tropics,* Institute of Biology Series, Studies in Biology No. 58, London.

Mani, M. S. (1974) *Ecology and Biogeography of India*, Dr W. Junk bv, Publishers, the Hague.

Matthew, K. M. (1983) *The Flora of the Tamil Nadu Carnatic* (illustrated), Rapinat Herbarium, St Joseph's College, Tiruchirapalli, Tamil Nadu.

Santapu, H. & Henry, A. N. (1983) *A Dictionary of Flowering Plants in India*, CSIR, New Delhi.

The Concise Oxford Dictionary of Botany (1992), Oxford University Press, New York.

The Wildlife (Protection) Act, 1972 (1994) Natraj Publishers, Dehra Dun.

2 Bringing the Forest to the Village

The Creation of Chakrashila Wildlife Sanctuary in Assam

NATURE'S BECKON

India

This account from Assam in northeastern India describes how a small NGO, Nature's Beckon, worked with tribal (*adivasi*) villagers to protect a 4,500-hectare area of forest from timber traders and game poachers. Their struggle over more than a decade resulted in the creation in 1994 of the Chakrashila Wildlife Sanctuary,[1] a legally protected conservation area that is managed by local villagers with the full support of the state and central governments. Chakrashila is one of only two known areas in India where the Golden Langur monkey (*Presbytis geei*) still lives in the wild.

Assam is one of the most important areas in the northeast in terms of biodiversity and richness in species, varieties and endemics. According to the *State of Forest Report* (Government of Assam 1993), 31 per cent of the total geographical area of Assam is under dense or open forest cover. This shows a decrease of about 0.31 per cent from the figures for the year 1991, possibly due to the loss of forest cover to agricultural and development purposes. Most forests in India are surrounded or inhabited by *adivasi*, who for centuries have lived in harmony with the forests. Severe external pressures, however, especially in Assam, have led inevitably to the loss of flora and fauna over a period of time. More than 70 per cent of the Assamese population depend on agriculture and forest resources as their main source of livelihood, although only 20 per cent of the geographical area of the state has been utilised for agriculture. Where shifting cultivation or *jhuming* forms a part of traditional land use patterns, deforestation has been inevitable. With raw material in increasing demand for industrial purposes such as paper manufacturing and railway construction and maintenance, forests are fast disappearing. Nevertheless, Assam is famous for its Kaziranga National Park and Manas Wildlife Sanctuary, which have their own unique biodiversity. There are eight other sanctuaries, and the Chakrashila Wildlife Sanctuary is the latest addition to this list.

58

Golden Langur

It is almost certain that the range of the Golden Langur's habitat is within Indian territory and confined to the forests of Assam, making this state the exclusive guardian of this endangered species for the rest of the world. The Golden Langur was discovered in 1953 by the naturalist Edward Pristichard Gee, then secretary of the Eastern Region of the Indian Board of Wildlife, in a forest patch on the Assam–Bhutan border, at an altitude of 2,400 metres and between the rivers Sankosh and Manas. It was assumed that the presence of these animals was exclusive to this region. No further sighting of this shy and arboreal animal was reported until 1987, when members of Nature's Beckon, out on a birdwatching tour, sighted Golden Langur in the forests of Chakrashila Hills. The rediscovery of these monkeys, initially thought to have migrated to Bhutan from Assam, was kept a closely guarded secret by members of Nature's Beckon for fear of alerting poachers.

The Golden Langur, once thought to be extinct in India, is now listed under Schedule 1 of the Wildlife Protection Act of India, 1972 and the Red Data Book of the International Union for the Conservation of Nature (IUCN). This golden-haired monkey is a herbivore, and is classified under the sub-family *Colobinea*. Primates have long been an integral part of Hindu mythology, with stories of the ape god Hanuman dating from the Ramayan and Mahabharata periods. Even today, Hanuman is worshipped as a symbol of strength. Ironically, like so many other animals, the Langurs face the threat of extinction at the hands of the very species which so reveres them.

The Chakrashila Wildlife Sanctuary is now recognised as the natural habitat of the Golden Langur by the Zoological Survey of India (ZSI), The School of Fundamental Research, the World Wildlife Fund (WWF) and the Department of Forests, Assam. The protection of this vital forest is a tribute to the combined efforts of local people and concerned activists.

Regional Background

The northeastern region presents a diverse system of habitats ranging from tropical rainforests to alpine meadows. It is largely a humid tropical region with two periods of rainfall; the winter rains come from the west and the summer rains are brought by the monsoon winds. In eastern Himalayas, the rainfall ranges from 125 to 500 centimetres; in Assam from 178 to 305 centimetres. This uneven distribution affects the region in two opposite ways: floods and droughts. The temperature in the region varies with location, elevation, topography, rainfall and humidity. The winter temperature in Shillong, for example, varies from 4°C to 24°C; in Gangtok, from

9°C to 23°C. The summer temperature in the region rises to 40°C.

The Chakrashila Hills are situated in Dhubri District of western Assam. The forest is inhabited by tribal communities and, until recently, had the official status of reserved forest. Encouraged by the negligence of state government agencies, however, firewood and charcoal merchants thrived, while hunting was the common pastime of people from all walks of life. Villagers of the region frequently worked as labourers for the timber merchants. They were thus unwittingly destroying the forests on which they depended, and losing their only source of fuel, food, medicine and building materials. Sal (*Shorea robusta*) and other valuable trees were felled and smuggled out daily. Unchecked logging left more than five square kilometres of forest along the periphery of Chakrashila completely denuded, which in turn forced villagers to move deeper into the forest, causing drastic depletion of the remaining vegetation.

This was the unfortunate situation in the forest reserve until the early 1980s, when a voluntary organisation of nature lovers, Nature's Beckon, became aware of the problems in Chakrashila. Some members of the group had frequented the area in their youth. Returning now for birdwatching trips, they were dismayed at the miserable state of the forest and resolved to find a way to save what remained of Chakrashila.

Chakrashila Forest has several villages on its periphery, inhabited mainly by the Bodo and Rava tribal peoples. Many *adivasi* villagers had been hired to log, but were severely exploited by the timber traders and poachers and remained very poor. Although they realised they were being cheated out of their fair share by the merchants, they did not retaliate. Lacking self-confidence, leadership and unity, they tended to be passive in the face of a deteriorating situation. Nevertheless, they possessed a great wealth of knowledge about their forests, and showed an enormous commitment to protecting these ecosystems once they felt sure they had external support for their cause.

Nature's Beckon realised that the tribal people were knowledgeable about the resources and their sustainable management. We also came to understand that the conservation of Chakrashila would be impossible without their cooperation. To begin with, people would have to stop being tools of the exploiters. This called for an action plan to educate people about their own welfare and to motivate them for the causes of conservation. But no meaningful education can be imparted by force or by arguments. A congenial relation is essential between the leader and the people. The leader must enjoy the full confidence, love and respect of the people. All this demanded patience, perseverance and an ability to keep calm in the face of adversity.

Working with the People

The office of Nature's Beckon is situated in the town of Dhubri. As communication between Dhubri and Chakrashila is poor, a temporary hut was built at Jon-nagra Village on the periphery of the forest. The hut was strategically located next to the house of Longthu Rava, a villager and friend since our childhood days. Longthu introduced us to the local youth and village elders, and acted as a guide in the region. Attracted by our interest in birdwatching, trekking and identifying birds and animals, a number of young people accompanied us on our trips. In time, they asked to be enrolled as members of Nature's Beckon. We presented them with the organisation's badge which gave them great pleasure.

Most of the time that Nature's Beckon spent at Jon-nagra was used preparing checklists of the fauna and flora of the region. We spent much of our time in the forests studying plant and animal ecology. Gradually local people realised our intentions and began to frequent our hut. In the evenings, over an open fire, we would discuss various aspects of the environment and the interdependence of plants and animals. In due course discussions with the villagers turned to the problems of Chakrashila, and people became convinced that the forest could not be protected unless they acted themselves. A number of young *adivasi* showed a particularly keen awareness of their problems. They agreed that the circumstances of their lives were increasingly difficult owing to the loss of forest resources on which they depended. But they did not know how to prevent the influential merchants and poachers from destroying the forest.

We explained that smuggling timber and poaching were illegal activities, punishable by law, and that those who broke the law had no moral ground from which to fight us, provided that the people took a firm stand against them. Although our younger supporters were enthusiastic and ready for immediate action, other members of Nature's Beckon were aware that the time was not ripe for such a stand. We felt that before any drastic steps were taken against the traders and poachers, it was necessary to gain the full support of the whole community. It was important that everyone understood and endorsed the intentions of Nature's Beckon, and that people were convinced they would benefit from actions taken against illegal exploiters of their forest.

During the next phase of our work, we visited every household in the village of Jon-nagra. We discussed the problem of forest exploitation with the elders of each family and explained how gradually other difficulties that the villagers face, in the areas of health, education and poverty, might be overcome by setting up sustainable development projects. During these talks we emphasised the importance of women's involvement in environmental management and development of the region.

It took about a year to mobilise the support of the villagers and to prevent the timber smugglers from entering the forests of Chakrashila. In the month of November, when the harvest is complete and people have suffi-cient food and leisure time, we began to act. In our first physical confronta-tions with the timber smugglers, some of the young *adivasi* were injured. But we lodged no complaint with the police or the Forest Department, as this would have made the villagers dependent on these officials.

Having violated the law, our opponents could not lodge complaints against the youth. Instead, the smugglers tried their best to turn the police and the Forest Department against the members of Nature's Beckon. But we continued to detect the movements of smugglers and to chase them away from the area, without asking for help from the Forest Department. We also confiscated the saws, ropes, chains and other implements used in tree felling and handed them over to the Forest Department. At last the officials began to appreciate our genuine efforts to save the forests of Chakrashila. According to then Divisional Forest Officer in Dhubri, D. Jaman: 'It is only with the help of such an organisation that the Forest Department will be able to achieve its *prima facie* [sic] objective of preserva-tion and improvement of forests.'

On one occasion, the villagers confronted and drove away a large gang of timber smugglers. In appreciation, the State's Principal Chief Con-servator of Forests, R. N. Hazarika, wrote to us:

> Your organisation has done a commendable job in detecting timber smugglers and in the seizure of saws used in the illegal felling of trees in Chakrashila Forest, for which I congratulate you and the other members of Nature's Beckon and offer thanks for such bravery in conservation and protection efforts. I hope that in future your organisation will continue to extend such services for the cause of the environment and the preservation and protection of wildlife in the State.

On another occasion the villagers set fire to a truck that had entered the forest to carry away the smuggled timber. After this incident the smugglers dared not bring a truck into the forest again. In a third instance, villagers, armed with their bows and arrows, surrounded a gang of poachers inside the forest. They seized four guns and ammunition, and handed them over to the Deputy Commissioner of Dhubri, Kanak Sharma. In recognition of the dedication of the villagers and members of Nature's Beckon, the state government awarded them a grant of Rs 5000 from the Chief Minister's Relief Fund. This encouraged the villagers and further raised their morale.

Although the battle against smuggling continued intermittently, the villagers did not lose their courage and zeal. But we realised that it was impossible to do any constructive work in such a hostile environment. We were eager to continue our work in a peaceful manner in Chakrashila. The discovery of the Golden Langur in the region, as well as our prolonged association with the villagers, strengthened our hold over the management of the Chakrashila Forest.

Bringing the Forest to the Village

The external factors that disturbed our progress subsided to a large extent. The dominance of *sal* in the periphery of the forest facilitated the regeneration of many plant species, some of which are used for medicinal purposes. And villagers oversaw the regeneration of Chakrashila Forest with round-the-clock surveillance.

At Nature's Beckon, our concern was not forest conservation alone, however, but also the well-being of Chakrashila villagers. We knew that the protection and sustainable development of the region depended on local people, and that given the paucity of funds, our efforts towards economic upliftment had to be based on the management and development of forest resources. By gathering and selling minor forest products such as bamboo or grasses for thatch and fodder we could earn some revenue to support the villagers. We encouraged people to cultivate their traditional food crops and to continue gathering secondary products like snails, rats and crabs.

To improve the availability of food we helped develop kitchen gardens in the village, supplying people with various vegetable seeds, and initiated poultry and pig farms. The results of these efforts not only sustained people, but also provided them with enough surplus to earn extra money. Some *adivasi* families began weaving, which proved to be another important source of income. These small but significant achievements helped to strengthen people's sense of identification with the forest.

Our work in Jon-nagra encouraged people of other villages – Abhyakuti, Bandarpara, Kaljani, Damodarpur, Banshbari – to join our organisation. As our goals widened, we felt the need to have a central office for the conservation and development of Chakrashila. We renovated our small hut and created more space, planted trees around the structure and sank a well. In time it expanded into a training unit, and came to be known as 'Tapovan'. This became an important community centre and helped to build cohesiveness among the different villages.

Currently, Tapovan is a vital education centre for women, young people and children in Chakrashila, and also offers hospitality to naturalists and interested tourists from outside the region. This has put the villages of Chakrashila on the map, and local people are motivated to do more for the protection of their region. It is perhaps worthwhile to mention that Tapovan's creation was at no extraordinary cost – most of the resources for its construction and labour came spontaneously from local people.

A popular slogan used in the early days of the development of Chakrashila was: 'Bring the forest to your village'. To decrease the pressure on the surrounding forest the villagers were encouraged to plant various species for medicine, food, fuel and thatch in their villages. The peripheral zone was used for plantations of fuel and fodder species that fulfilled local requirements. Furrows were dug connecting cultivable lands of Jon-nagra to a perennial spring, Mauria Jhora, which radically decreased people's dependence on the forest.

While working for the betterment of Chakrashila we conducted a survey of the forest that included checklists of the flora and fauna. This had never been done by the Forest Department. To the surprise of many department officials and naturalists, it was discovered that, in addition to the Golden Langur, there were many other endangered plant and animal species in the region. After the documentation of these facts it seemed logical that the conservation status of this unique hill reserve should be upgraded to that of wildlife sanctuary.

Conclusion

Permanent protection of Chakrashila will benefit villagers living on the periphery directly. If the forest cover remains intact, the perennial water sources (which are used for household purposes and for cultivation) will not dry up, and biological wastes from the forests will enrich the soils continually. Legal designation as a wildlife sanctuary will also contribute to the socio-economic development of these villages.

Over the past 12 years, the people of Chakrashila have sacrificed much to protect the forest, often putting their own safety at risk. It would be an injustice if, ignoring these efforts of the people, the forests were allowed to slip into the hands of timber merchants or unscrupulous politicians. Nature's Beckon put these views forward to the state government. When we received no response, we began to appeal to the public to put pressure on the government. We were supported by various NGOs and some prominent personalities such as Dr Ashish Ghosh, Director of the Zoological Survey of India, Romulus Whitaker, Director of the Centre of Herpetoloy of the Madras Crocodile Bank Trust, and Dr Nandita Krishnan, Director of the CPR Environment Education Centre in Madras. They personally appealed to the Chief Minister of Assam and to the Ministry of Environment and Forests, endorsing our demand to upgrade the status of the Chakrashila Hill Reserve.

During this period of lobbying and dialogue we invested our time in the management of the Chakrashila Forest. We planted trees in the denuded areas, prevented poachers from entering the area and continued to check timber smuggling. We also planted fruit trees useful to the Golden Langur and other animals, checked the indiscriminate burning of the forest and provided artificial salt licks for the large mammals. The villagers volunteered to clear weeds and guard the forest from any abuse. Signboards for the preservation of wildlife were put up at various points.

We managed eventually to gather support for our cause from the media and the broader public of Assam. The issue was debated in the State Assembly and finally, on 14 July 1994, the Governor of Assam upgraded the former hill reserve into Chakrashila Wildlife Sanctuary. Since then our responsibility has increased, as we now have to show that the sanctuary can be managed best by the villagers and Nature's Beckon, with the benevolent support of the government of Assam.

NOTES

1 The wildlife sanctuary is located between the latitudes 26°15' and 26°26' north and between the longitudes 90°15' and 90°21' east.
2 Through notification number FRW-6/93/63, under section 26A (1) (b) of the 1972 Wildlife Protection Act.

BIBLIOGRAPHY

Ali, Salim, (1977) *Field Guide to the Birds of Eastern Himalayas*, Oxford University Press, New Delhi.

Choudhury, D. (1994) *Hima-Paryavaran*, Vol. 6, No. I (June), G. B. Pant Institute of Himalayan Environment and Development.

Daniel, J. C. (1983) *The Book of Indian Reptiles*, Bombay Natural History Society, Bombay..

Datta, Soumyadeep (1995) *Birds of Dhubri District. A Check List with Trip List*, Nature's Beckon.

Datta, Soumyadeep (1995) *Conservation of Biodiversity and the Role of NGO (Banbani)*, Nature's Beckon.

Datta Soumyadeep (1993) 'Nature's Beckon Regreening the Hill of Assam', *Debacle*, Vol. 7, No.4.

Gosh, T. K. (1993) *Field Discovery*, Indian Quarterly, January–March 1993.

Government of Assam (1994) *State of Forest Report*, Department of Forest, Assam.

Jacob, Sunil (1995) 'A Home for the Golden Langur', *News Environmental Education*, Vol. 1, No. 4 (September), Centre for Environmental Education.

Prater, S. H. (1988) *The Book of Indian Animals*, Bombay Natural History Society.

3 Saving the Western Ghats
Blueprint Solutions versus Villagers' Rights and Responsibilities

PANDURANG HEGDE and S. T. SOMASEKHARE REDDY
Prakruthi, India

Prior to the British entry into India, the forests were considered the common property of neighbouring communities. The dependence of local people on the forest was mutual. The forest was closely integrated into the farming and village industry system, and this system required that people protect the forest in order to survive. As the British attempted to extract more wealth from India, they claimed sovereignty over the forests and made laws to enclose them. This enclosure had a disastrous effect on local villagers. It reduced their agricultural productivity, damaged their village industries, denied them access to wild foods and fodder, harmed their health and had a multitude of other impoverishing effects.

The Forest Department saw its role primarily in terms of maximising revenue for the state. Income was dependent mainly on hardwood timbers, in high demand in Western industrial and urban society. People who had lived on numerous forest products sustainably extracted for centuries were now forced to sell headloads of fuelwood or smuggle out felled timber, while clearing further pieces of forested land for cultivation – all this simply to survive. The enclosure of the forest also caused ecological damage. Commercially valuable species, such as teak and rosewood, were felled selectively. Where reforestation was undertaken, it was only with a very limited number of species. This constituted a considerable loss in bio-diversity and in the sustainability of the ecosystems.

After the achievement of national independence, this system of forest management was continued by the new Indian state. During the decades that followed, the commercialisation of forests and their exploitation only intensified. Subsequent forest laws introduced more rules and powers enabling the state forest services to curtail the user rights of local people. This has been the general picture all over India. Forest management in the State of Karnataka has followed the same pattern. Bilateral and multilateral development agencies, such as the World Bank, the Asian Development

Bank (ADB) and British Overseas Development Assistance (ODA), as well as foreign consultants, have contributed to further entrenching the current system. Substantial funding and loans have been invested in the state forest bureaucracies and massive extension of monocultural plantations, notably under the pretext of 'social forestry'. Over the last 15 years, these interventions have met with vehement opposition from millions of villagers whose commons have been appropriated for monocultural tree planting. Without any evaluation or proper reflection on the pros and cons of these so-called 'development schemes', however, new concepts such as 'joint forest management' have been introduced.

At the beginning of the 1990s, the Karnataka state government requested British ODA[1] to fund a forestry project in the Western Ghats, a forested mountain range on India's west coast. The result was the Western Ghats Forestry Project, designed by British scientists and funded by ODA. The project aims to enhance the management capacity of the Karnataka State Forest Department, and in particular to enable it to respond to conflicting demands from different users for access to the forest. Yet it is the Department and its forerunner, the colonial forest service, that are chiefly responsible for the current state of affairs: loss of natural forest and increasing conflicts. The Karnataka State Forest Department shows no signs of a major shift of attitude towards the needs and role of local villagers. It is therefore to be feared that improving the Department's management capacity will only exacerbate these problems. The ODA scheme requires that a huge bureaucracy, with extensive infrastructural facilities, be developed. Once ODA has withdrawn, however, these will need to be funded by the state. The latter will almost certainly raise additional revenue from expanded forest exploitation. The infrastructure, which includes the construction of new roads into the forest, will make the forest more accessible to official exploitation, as well as smuggling.

The Western Ghats Forest Project also means to ensure that poor people, women, tribal people (*adivasi*) and other disadvantaged groups who are dependent on the forest 'are not worse, and preferably better off'. Unfortunately, so far the project has caused only hardship for local villagers – particularly the already marginalised, those with little or no land. The ODA-funded plantations have been mainly on village commons, which once provided pasture, fuel, manure, medicinal plants and other products to meet basic needs. Families have been evicted from cultivated lands on which they have 'encroached', depriving them of their livelihood without any compensation. Women are most affected by the decrease in availability of fuel with which to cook. They must go longer distances in search of fuelwood. A reduction in the availability of fuel could result in insufficiently cooked food, with a reduction in its nutritional value and hence serious damage to the health of families. The *adivasi* affected in this area are the

Gowlis, originally pastoralists. They had earlier been forced to settle in villages, having to reduce their herds since there was insufficient grazing land. Their situation has now worsened with the implementation of the British project. Impoverished villagers are forced to take more from the forest for bare survival, leading to greater forest destruction.

Another objective of the ODA forest project is minimisation of forest loss. In reality, however, the project encourages the further conversion of natural forest to monocultural plantations. Experience has also shown that certain tracts of forest are preserved, only to be felled at a later stage if this proves lucrative. Sustainability merely means ensuring future revenue flows from the forest, not ensuring that the forests remain. According to the project, gaps in the forest will be planted with 'valuable timber species', with the aim of restoring the natural mix of the forest. Ironically, several of these gaps were created by the Forest Department, which clearfelled existing natural forest or plantations. In all these gaps, considerable natural regeneration was already taking place. The Department's selection of species can never restore the natural mix of forest, which normally contains several hundred different plants and trees. Since the start of the project, there is absolutely no evidence to show that any attempt has been made or is proposed to preserve or restore the area's biodiversity. In 90 per cent of the plantations established in the majority of villages surveyed, *Acacia auriculiformis* (an exotic imported from Australia) has been planted. During meetings with the department, villagers have asked in vain for a reduction of the number of acacia plants.

The project aims, again, to increase the understanding of the Western Ghats ecosystem. But the major point that is not understood is that people and cattle form an integral part of the forest ecosystem. Preserving the system involves reducing intervention and leaving the system to evolve in its own time and in its own manner. There is a clear conflict among these various aims of maximising the income of the Karnataka State Forest Department, increasing the welfare of the people, and conserving the forests. Villagers have been managing their surrounding forests for centuries, obtaining many of their essential needs from them. Some of these traditional forest management systems still exist, and it is these that need to be studied, consolidated and promoted as a concrete and inspiring example for other villages. To counter the negative effects of the official forest project and similar interventions, and to demonstrate a more viable alternative, a number of such in-depth case studies were undertaken. Below is an analysis of a traditional village-based forest management system in the Western Ghats. It survived both the colonial era and post-independence administrative reforms, and today represents a straightforward and viable social institution, controlled by and for the benefit of the villagers themselves.[2]

Introduction to Halikar Village

On India's southwest coast, wedged between the sea and the Western Ghats[3] mountain range, we find the village of Halikar, which falls under the administrative jurisdiction of Uttara Kannada District in the State of Karnataka. The Western Ghats contain an almost unbroken chain of ever-green tropical forest stretching from the south of India to the hills of Goa. Further north the forests are smaller and become deciduous. This change is related to climatic conditions. For example, according to Pascal (1996), the dry season in Kerala is only one to two months, whereas it exceeds seven months in Bombay. Another factor is the gradient. Pascal mentions, in this respect, the phenomenon that the region of Agumbe, one of the wettest places in India, receives over 7,000 millimetres of rainfall annually, whilst 50 kilometres inland the rainfall drops to only 200 millimetres. Hence the structure and composition of the forest changes. 'The formation of elegant trees with straight trunks, their crowns emerging at 35–40 metres, sometimes dominated by giant emergents at 50–60 metres, rapidly makes room for a low forest popularly called *chola*, composed of small trees 15–20 metres in height, crooked, low-branched and heavily covered with epiphytes' (Pascal 1996). The warm and moist climatic conditions of the Western Ghats are, in most places, accompanied by soils which are extremely vulnerable to impoverishment and erosion. In the regions extending between the foothills of the Ghats and the sea human impacts have caused the replacement of evergreen forest by deciduous forest. Only under conditions of strict protection have some evergreen species survived in the deciduous canopy (Pascal 1996).

Halikar is surrounded on its eastern and northeastern sides by the forest area that goes by the same name. Halikar Forest measures about 245 acres. On the western side, beyond the paddy[4] fields and the backwaters,[5] is the Arabian Sea. In the south Halikar is surrounded by the villages of Hegade, Vakkanahalli, Chitragi and Masuoor, and by the town of Kumata. A road connecting Kumata and Masuoor marks the boundary of Halikar Village on its northeastern side.

Halikar Forest is deciduous, with a few evergreen species. This area receives 3,000 mm in rainfall per year, most of it during the monsoon in the months of April–May and September–October. In the dry months the temperature goes up to 43°C and the lowest temperature in winter (October–February) is 18°C. Formally, the Forest Department has juris-diction over Halikar Forest. The wildlife in this area includes jackal, wild cat, boar and rabbits. Soil conditions range from sedentary laterite in the forest area to sandy soils towards the coast and clayey soils near the tank[6] and the agricultural fields.

The caste groups

In order to understand the way in which Halikar Forest is used and managed by the villagers, it is necessary to understand their social and economic conditions. As is typical of most Indian villages, Halikar's population is subdivided on the basis of caste groups. Castes in India are hereditary classes that correspond with traditional occupations. As such, caste groups determine all aspects of social and economic life in Halikar Village, and the caste organisations make their members adhere to rules concerning the use of the forest.

There are eleven caste groups in Halikar. The biggest group is the *Harikanta* (fishers) group with 60 households, followed by *Patagara* (39 families) and *Gunuga* (31 families). The other castes are listed in Table 3.1. Normally a person will take up the traditional occupation (such as priest, carpenter, or blacksmith) linked to the caste, unless he or she opts for another profession, if such an opportunity arises. Employment outside the prescribed profession brings an assured additional income and may enhance the social and economic status of the family, and even that of his/her entire caste group in the village.

Table 3.1   Caste groups in Halikar

Caste group	Number of families	Traditional occupation	New occupation
Achari	4	carpenter	
Kodiya	2	barber	
Patagara	38	stone cutter	driver (2), business (1)
Madivala	7	washerman	
Muhkri	18	agricultural labourer	government service (2)
Nayak	1	driver (1)	
Gunuga	31	pottery	
Shanubog	5	business	business (1)
Havyaka	12	priest	teacher (3), retired teacher(3)
Hulsvara	3	pottery	driver (1), bank employee (1), business (1)
Harikanta	60	fisherman	government service (1), business (1)

The main occupation of all the caste groups, except for the fishermen (*Harikanta*), is agriculture. Of the 182 families who constitute Halikar Village, 129 own agricultural land. Non-*Harikanta* landless families still depend on agriculture: many work on the land as labourers. The amount of land owned by each caste gives an idea of its economic strength (Table 3.2).

Table 3.2   Caste groups and land ownership in Halikar

Caste	Number of families	Landless	< 2 acres	>2 acres
Achari	4	–	4	–
Kodiya	2	1	1	–
Patagara	38	3	31	4
Madivala	7	–	7	–
Mukhri	18	2	13	3
Nayak	1	–	1	–
Gunuga	31	5	25	1
Shanubog	5	3	2	–
Havyaka	12	1	2	9
Harikanta	60	37	23	–
Total	178	52	109	17

An examination of the landownership pattern shows an unequal distribution of land among the caste groups. Most *Havyaka* (Brahmin) families have substantial landholdings, whereas only a few families belonging to the other caste groups own more than two acres. This tells us that the *Havyakas*, who are already high on the social hierarchy, also enjoy greater economic status. The fisher community has the largest number of landless. If few landless work as agricultural labourers, this is because most landless earn a living from fishing.

The land-use pattern in Halikar Village has undergone a striking change in three decades. Until the 1970s, the village was renowned for the cultivation of sugar cane and the preparation of *jaggery* (sugar syrup). This tradition came to an end because of a scarcity of fuelwood to process the cane into *jaggery*. Another abandoned tradition is the cultivation of a second crop of paddy during the dry season with irrigation water from the tank. This practice has ended because the tank has silted up.

In spite of the stratification of Halikar Village into different castes, each of which has distinct customs, a strong network of social organisations binds the caste groups together and maintains cohesion. Each caste group has its own social organisation, which carries the name of the deity associated with the traditional occupation of that caste. Each organisation puts on an annual festival in the village in honour of its deity. These organisations have responsibility for resolving conflicts among their members. In addition, they administer a small credit fund to assist members of the caste. Although these funds are relatively small, they are held in high esteem as the fund carries the name of the caste's deity.[7]

Besides the caste organisations, Halikar has its *Van Panchayat*, or Village

Forest Committee, and the Youth Organisation. The Village Forest Committee came into being in 1924, at the initiative of the British colonial government, and consists of villagers who are elected to represent their respective castes. The Committee is chaired by an elected president. Its main responsibility is the management and preservation of Halikar Forest. Initially, the secretarial post was also honorary. Since the responsibilities of the Committee and the consequent workload of the secretary have expanded, however, a salaried secretary is now appointed. The Committee maintains intensive contact with the Forest Department and the court.

The Youth Organisation was established in 1961 by young people in the village to assist the Forest Committee in a dispute with the government concerning control over Halikar Forest. This involved the preparation of litigation, as well as practical protection measures to maintain the forest's integrity. The organisation later took responsibility for various social and economic activities, such as improving the local school.

The Village Forest Committee

The evolution of the Village Forest Committee in Halikar is closely related to the history of British colonial forest policy, and to the so-called Indian Forest Act of 1924. This law prescribed that the government should transfer management and user rights over forest adjacent to villages to local committees. Those village communities which wished to assume such management responsibilities were in turn obliged to form a committee consisting of elected inhabitants of the village. They were also supposed to adhere to regulations formulated by the government.

The year the law was enacted, some 50 families from Halikar formed an executive committee consisting of the elected representatives of nine different caste groups. They accepted the condition that once in every three years they would hold elections and that the membership list would be submitted to the *Tashildar* (head of the government administration of the *Taluk*, or lowest administrative unit). It was also accepted that the *Tashildar* would preside over the elections, with the power to cast a vote when candidates failed to obtain a majority. After the Village Forest Committee was established, an area of 225 acres of forest surrounding the village was demarcated and placed under its control.

The Village Forest Committee appointed an honorary secretary and a salaried watchman. (Ever since, a secretary has handled the daily affairs and executed the decisions of the Committee). The Committee also formulated a policy to preserve Halikar Forest and started to regulate the collection of forest produce and cattle grazing. It issued a pass to every household in the village, which gives permission to graze cattle in the forest and to collect a specified amount of forest produce, such as fodder and firewood. Each year these passes were to be renewed, with payment of a fee. Those who exploited the forest without a pass were liable to be fined by the Committee. The collection of forest produce was subject to strict guidelines. Villagers were not supposed to cut green leaves from plants below two feet in height, branches were not to be more than two feet long, and forest products were not to be sold outside the village. Whoever reported a violation of the above rules was rewarded with 50 per cent of the fine. To mark the boundaries of the forest and to prevent encroachment, all households were summoned to erect a fence around the forest and to maintain it. Villagers were also obliged to participate in tree planting and to contribute saplings whose species were selected by the committee. Any absentee was likely to be fined.

These regulations are still applied and, in fact, have become stricter. Yet when we examined the records of the Village Forest Committee it appeared that, over the past 70 years, a meagre amount was collected through these fines. The villagers explained that those who are fined face stigmatization by the rest of the village, a far greater punishment than the fine itself. At the same time, villagers whose efforts benefited the forest and the Committee's work are generously rewarded. For example, a previous president of the Committee, who was rewarded 10 rupees in appreciation of his work, is still honoured by the other villagers, who added the name of the reward to his surname. The effectiveness of the Committee is also reflected in the absence of any external official intervention (with the exception of an annual official review of accounts and the supervision of elections every three years) until 1979.

Disputes over Halikar Forest.

In 1957 independent India reorganised its provincial structure on a linguistic basis. Uttara Kannada District, which formerly belonged to the Bombay Province, became part of the new State of Karnataka, where the majority speak Kannada. In 1960 the government of Karnataka introduced new forestry legislation which gave the Forest Department monopoly powers over the management of the state's forests. It thereby annulled the rights and responsibilities of all other institutions. The dramatic effect of this ruling can not be overstated. The policy effectively killed a refined mosaic of local institutions: the *Patel* (village headman), *Shanubog* (village secretary), and the Temple Forest Committee, which was responsible for controlling access to forest resources during colonial and pre-colonial times.

Likewise, this policy seriously threatened the villagers' control over Halikar Forest. In 1979 the District Commissioner passed an order to the effect that the Village Forest Committee should surrender its authority to the Forest Department. The villagers decided to challenge this order in the high court, where it dragged on for almost a decade. In 1989 the high court judgement supported the villagers' claim: the government could not abolish a right that the villagers had enjoyed since 1924 when it was formally granted by the colonial government.

This was only the first of a series of assaults on the forest by outsiders which the villagers opposed. During the 1980s the government passed an order for Halikar Village to release 50 acres of forest to be converted into building sites to accommodate the expansion of the nearby town of Kumata. When the villagers challenged this policy the order was withdrawn. When the Konkan railway line, connecting major towns on the west coast, was constructed, Halikar Village had to sacrifice 14 acres of its forest. The railway track was laid without any prior notice to the Forest Committee or the villagers. In fact, compensation for the loss of forest was pocketed by the Forest Department. When the Committee finally succeeded in claiming compensation, the Forest Department only agreed to pay for the loss of trees, not for the loss of land.

Soon after, the Committee faced another battle with the Ministry of Power, which (again without asking permission) cut down a large number of trees in order to run a powerline through the forest. The villagers protested against this encroachment and the Committee put forward a claim for compensation. Although the government ministry turned a deaf ear to these protests, the case demonstrated once again the vigilance of the villagers. In a separate incident, the then *Taluk* Board president had plans to cultivate pineapple and *chikoo* fruit in Halikar Forest. The villagers vehemently opposed this idea, and the plan had to be withdrawn.

One of the most far-reaching interventions, however, took place more recently with the introduction of the 'social forestry' scheme by the Forest Department. Although the Village Forest Committee requested that the Department plant a selection of endemic trees, the latter choose to plant mainly acacia. Despite the Committee's objections, the Forest Department continued to plant acacia during the years that followed. In addition, the Department fenced off the forest to prevent cattle grazing. Not surprisingly, the owners of cattle complained about the resulting shortage of grazing land. Other villagers argued that it would enable the forest to regenerate, but this opinion was countered with the argument that the remaining grazing land is now bearing the full brunt and will soon suffer serious erosion. This discussion continues.

Villagers perceive that a key threat to the forest is the government's failure to conduct elections for the Village Forest Committee. The situation became acute in 1993, when the term of the Committee members came to an end and they had no formal decision-making powers. The *Tashildar*, who was supposed to arrange the elections, turned a deaf ear to continued requests by the villagers to perform his task. The issue was picked up by the local press. As a last resort, the villagers asked the District Commissioner to intervene. The latter ordered the *Tashildar* to organise the elections, and a new Committee was finally elected for a three-year term.

Regulation of Forest Use

The demand for fuelwood places great pressure on Halikar Forest. The number of fuelwood uses depends very much on the cultural codes that determine the day-to-day life of the different castes. The *Havyakas*, who form a priestly caste, are bound to the religious dictum that they have to worship God after bathing in the morning. Consequently, the *Havyakas* use substantially more firewood for bathing purposes than most other castes. On the other hand, since most *Havyakas* are relatively well off, they can afford to purchase gas for cooking purposes.

Over the years, the Halikar Village Forest Committee introduced more restrictions on the use of the forest. No villager is allowed to cut living or dead trees. One may collect fallen twigs, but only if one has obtained a pass book from the Committee. Dead or fallen trees are auctioned by the Committee, and every villager is eligible to participate. The Forest Department runs a fuelwood depot 20 kilometres away from the village. Again, only if villagers obtain a pass can they buy a ration of 5 kilogrammes of firewood per card. To save time and transportation costs, villagers form informal groups whose members collect firewood from the depot on a rotational basis.

Over the last few decades the consumption of fuelwood has decreased due to changes in cropping patterns and the replacement of traditional pottery by modern materials. In the past, the production of *jaggery*, tiles and earthen vessels and the processing of *areca* nut and paddy required large amounts of firewood. In order to decrease the pressure on the forest and to avoid the high transportation costs of acquiring fuelwood from the outside, villagers made a conscious decision to stop *jaggery* production some 20 years ago.[8] With the arrival of a rice mill in a nearby village, which enabled the villagers to obtain semi-boiled rice, the inhabitants of Halikar also ceased to boil rice, a process which consumed large quantities of firewood. With the introduction of concrete roofs, and steel and aluminium vessels, the demand for pottery dropped. The families who continued the profession of pottery purchase their fuelwood from the government depot.

One advantage of the proximity of the town of Kumata is that it allows villagers – that is, those who can afford it – to purchase alternative sources of energy: kerosene and gas. Besides that, villagers have started to look for other sources of fuel, such as coconut trees. Some 45 families have grown coconut around their homes, and another 20 acres of farming land are planted with coconut trees. The owners of the trees can now meet their fuel requirements by using the dry leaves, pods and other residual material.

Leaf litter is widely used as green manure in Uttara Kannada, helping farmers to avoid leaching of the soil during heavy monsoon rains. In most cases, leaf litter is first used as bedding for cattle and then spread over the fields. In 1985 the Village Forest Committee prohibited the use of green branches for manure. The collection of dry leaves is regulated by a system of passes and quotas. Passes can be obtained from the secretary for a fee, and trespassing is punished by the Committee. Interestingly enough, the availability of chemical fertilisers had no impact on the demand for leafy manure, an indication that the farmers of Halikar place a higher value on the latter. They consider the application of fertilisers to be a waste as it washes away during the monsoon season.

The agricultural implements used by farmers are wood-based and long-lasting: wood is used for ploughs as well as for the handles of pick-axes and shovels. According to the farmers, Halikar Forest does not host the right sort of trees for these purposes, and people therefore call upon relatives from other villages or go to a neighbouring forest. The forest is not an important source of medicinal plants, either. Villagers have access to a hospital at close range. The forest contains a number of different fruit trees with nutritional value, such as cashew, *yamun*, mango and *kavali*. *Murgaloo* and *surgi* flowers are collected from the forest and used in the preparation of food. *Surgi* flowers are also auctioned to outsiders who use the flowers in the preparation of perfumes. Previously, the Committee put no restriction on the collection of fruits and flowers. Recently, however, it decided to auction all fruits, in order to generate more revenue. Since 1943, the Committee has promoted the planting of fruit trees. The Committee has no plans whatsoever to extract timber from the forest. The auctioning of dead or fallen trees brings in an average annual revenue of Rs 2,000.

Each family is expected to contribute labour to communal activities organised by the Committee, such as tree planting or the collection of fines. Should a family fail to make its contribution, it is entitled to explain its absence. If the Committee does not consider the explanation valid, it can levy a fine. In case of non-compliance, the Committee can prohibit the family entering the forest. Such a decision is made publicly known, and all other villagers are expected to ensure that the punishment is enforced. The family's rights are restored once the fine has been paid.

No serious conflicts have so far occurred between the formal bye-laws of the Committee and the customary rights of the villagers. The Committee has often preferred to take a flexible stand if it seemed necessary to accommodate pressing needs of residents. A serious clash of rights exists, however, if we consider the relation between the rights and responsibilities formally accorded to the Village Forest Committee, on one hand, and, on the other, the regulations by which the government attempts to overrule local decision making. Although the Village Forest Committee has been formally entrusted with control over Halikar Forest since 1924, the Forest Department did not bother to ask the Committee's permission when introducing social forestry. In much the same way, the Department erected a fence of barbed wire through the forest without consulting or even notifying the committee. The Department maintains that it is responsible for the management of forests all over the state.

Considering that the right to the forest is recognised on the basis of family rather than the individual, and that the father is the head of the family, women have limited space in which to play a significant role in forest management and conservation. Unless a woman is a widow, she cannot represent the family. We have not observed, as yet, that women opposed this situation. Neither the Committee nor the caste organisations are supposed to extend any favour to an outsider at the expense of the rights of the residents. For example, the sale of stones is bound by differential rates, whereby the residents pay lower prices than outsiders. A similar arrangement exists for the sale of minor forest produce. A new resident can only become a member of the Village Forest Committee if he has resided in the village continuously for ten years and seeks membership through one of the caste organisations.

The forest is open to all residents of Halikar, with no distinction made between ethnic groups in the village. Every villager is bound by the Committee's regulations, without exception. The Committee is accountable to the caste organisations and villagers are entitled to question the Committee's performance through caste representatives on the Committee. However, no villager can approach the Committee directly. Any grievance has first to be brought to the attention of the caste representative, who will bring it up in the Committee. This system has ensured the protection of the interests of all users of the forest since 1924, guaranteeing maximum transparency and equity.

The Future

The survival of the Village Forest Committee depends largely on how well it can adjust to rapidly changing social and economic circumstances. As we explained earlier, more and more villagers have started to relate to the outside world and leave the confines of the traditional occupation of their caste. As a consequence, it becomes more difficult for the castes to exercise control over their members' behaviour. The proximity of the town of Kumata is a major factor, as it offers new employment and influences lifestyles and expectations.

One tangible effect is that villagers become less dependent upon the forest for basic needs and income. The educated village elite has shifted to other sources of fuelwood and green manure, such as the Forest Department's wood and gas depot. Fewer women enter the forest to collect fuelwood. Caste groups, such as potters and carpenters, which traditionally used the forest to obtain their basic material, may soon cease to have such a direct interest in the forest as families shift towards other professions. Young people are increasingly looking outside the village for income-earning opportunities. The enthusiasm of the villagers, which over the past six decades sustained the forest and the Committee, may start to wane.

As the Village Forest Committee is constituted by representatives of the different castes, its strength very much depends on the vitality of the different caste organisations. Their performance, in practical terms, is measured by how dependably they assist members in times of economic uncertainty. These days, the credit systems of the caste organizations simply do not meet the members' need for cash. In that sense, the caste fund, which carries the name of the god the caste worships, becomes a mere conveyor of blessings. With the rising level of education, the sacred status attached to the fund may also vanish. These funds and the authority of the caste organisations may cease to provide social cohesion, thereby leaving room for dissension.

The Village Forest Committee has not addressed these problems. In order to ensure that the villagers of Halikar defend and manage their forest, it is vital that they continue to feel dependent on and responsible for the forest. It is also important that the Village Forest Committee maintains its authority. A number of concrete steps can be taken. The Committee could play a more active role in dealing with the economic problems of the villagers by exploring new employment opportunities. For example, it could promote the collection and marketing of non-timber forest products. In addition, the caste organisations could enlarge their credit facilities to accommodate the economic pursuits of members. The *Harikanta* (fishermen) caste organisation and the Village Forest Committee could assume a more active role in regulating the prawn fishing in the land adjacent to the backwaters. Only through strict measures can the degradation of this land be avoided.

Fortunately, the Village Forest Committee has a favourable record of having assisted in various village affairs. It was involved in the construction of the school and the youth club, and arranged the repair of the Dharma Shala – a simple guest house for travellers visiting Gokarna, a holy site. It helped in housing the *Anganawadim* (kindergarten). The Committee also helped to solve the acute shortage of drinking water for cattle by constructing water troughs. It helped fight cattle disease and arranged for substantial compensation when 400 cattle died. It is important that the villagers feel motivated to turn to the Village Forest Committee to deal with such problems. This is why the Committee should maintain its vibrancy, delivering solutions to such problems whenever possible.

The basic uses of the forest are to graze cattle and to collect fuelwood, leaf litter and minor forest products. It is essential that the Village Forest Committee solves the dispute over cattle grazing in the forest. As we explained earlier, the Forest Department closed the major part of the forest to grazing. Only 85 acres of forest land is now available to feed 227 head of cattle. As a result, cattle now start wandering through the village, destroying plants and vegetable gardens in villagers' backyards. The Village Forest Committee should address this problem to avoid further damage and to prevent a split in the village.

Currently, the collection of minor forest products is not supervised by the Village Forest Committee. The Committee could be more active, involving villagers as paid labourers to harvest non-timber forest products and thus creating extra income-earning opportunities. The committee should supervise the collection and sale of the products so that better prices can be negotiated and sustainability ensured.

The future of Halikar Forest also needs to be examined in the light of economic changes in the regional and national context. Uttara Kannada District is increasingly exposed to large-scale development schemes, especially in the power sector. There are at least nine large hydroelectric dams and one nuclear power plant under construction. There is also a steady increase in prawn and shrimp farming along the west coast. These developments constitute an enormous boost for the regional economy and will effect Halikar Village in one way or another.

NOTES

1 The Overseas Development Administration (ODA) is now the Department for International Development (DFID).
2 With thanks to Winin Pereira for his report 'The Western Ghats Forest Project', October 1995.
3 *Ghats* in Hindi means 'steps'.
4 Rice.
5 Creek separated from the sea by a narrow strip of land and communicating therewith by barred outlets.
6 Reservoir holding water, which is released by opening an outlet.
7 Accounts are normally maintained with help from a member of the *Havyaka* caste.
8 Villagers grew a total of 40 acres of sugar cane.

BIBLIOGRAPHY

Brandon, C. and Hommann, K., (1996) *The Costs of Inaction: Valuing the Economy-Wide Costs of Environmental Degradation in India*, The World Bank, cited in Anil Agarwal, 'Pay-offs to Progress', in *Down to Earth*, Vol. 5, No. 10 (October).
Pascal, J. P. (1996) 'Wild and Fragile', in *Down to Earth*, Vol. 5, No. 6 (August).
Pereira, W. (1995) 'The Western Ghats Forest Project', unpublished report, October.

4 Local Forest Management on the Frontier

Indigenous Communities Restore Their Forest in the Cordillera Mountains

MONTANOSA RESEARCH AND DEVELOPMENT CENTRE

Phillipines

Although the interplay of indigenous socio-political systems and the local communities' intimate understanding of the ecosystem has contributed to the maintenance of richly diverse forests in the Cordillera of the Philippines, the government continues to deny people's rights to the land and ignores their crucial role in the conservation and management of forests. The state has monopolised responsibility for protecting, managing and 'developing' the land as it pursues a dream of reaching industrialisation by allowing foreign entities to exploit the natural resources it has sworn to protect.

The Cordillera communities have to contend with deepening poverty and weakened local structures as well as foreign mining, logging, and infrastructural projects. Though thoroughly undermined by state forest laws, the indigenous systems still function today. This case study documents indigenous forest management systems in five mountain forest communities that are struggling to protect their forests and their customary land laws. A list of major regional flora and fauna is provided in Appendix 1.

The Five Study Areas

Three of the forest study areas – Demang, Bugang and Sisipitan – are in Sagada, one of the ten municipalities of Mountain Province in the central part of the Cordillera Administrative Region on Luzon Island. It is 150 kilometres away from Baguio City and can be reached by road. Sagada has a total land area of 8,568 hectares, 99.3 per cent of which is classified by the government as forest reserve. The climate is generally subtropical with two distinct seasons – the wet and the dry.

Bugang and Demang are between 1,500 and 1,700 metres above sea level, with forests dominated by the Benguet pine (*Pinus insularis*) interspersed with broad-leaved shrubs and small trees of various kinds. The ground is covered by grasses such as *Imperata cylindrica*, *Themeda gigantea* and *Miscanthus luzonensis*. The forests of Bugang and Demang show the results of tree-planting efforts by clans and families after the Second World War. Mount Sisipitan – the third area of study in Sagada – is still in a healthy state. Its vast forests serve as a watershed and provide hunting grounds for the Bontok, Kankanaey, Tingguian and Maeng tribes who dwell on the borders of Abra and Mountain provinces. The Sisipitan Forest has a wider range of elevation, from 1,400 to 2,200 metres above sea level. A large part of the middle portion is characterised by moss forest with patches of rainforest. There is pine on the Mountain Province side and dipterocarp on the Abra side. Rivers from Mount Sisipitan join the two major rivers in the Cordillera – the Chico and the Abra.

According to our surveys with local people, Sagada registered a population of 10,353 in 1990. These people belong to the Aplay tribe, a sub-group of the Kankanaeys, one of several ethnolinguistic groups in the region. The majority of the local population engage in rice farming and commercial vegetable production, raise poultry and livestock on a small scale, and work as seasonal wage earners. A few are employed by government and private agencies. Like other areas of the Cordillera, Sagada has retained many aspects of its indigenous socio-political system such as the *dap-ay*, a centre where community concerns and issues are actively discussed and resolved.

The other two forests that were studied are in Tubo municipality of Abra Province, which is on the western side of the Cordillera some 408 kilometres north of Manila and 197 kilometres northwest of Baguio City. Tubo is composed of ten *barangays* (the smallest administrative unit of the Philippine state) with a land area of 41,500 hectares, making it the second-largest municipality of Abra. The climate is moderately warm. The dry season extends from January to May, and the rainy season from June to December. Studies were carried out in Bana and Beew forests, located along the borders with Ilocos Sur and Mountain provinces. These are predominantly secondary rainforests with scattered stands of pine trees. The elevation of both areas ranges from 700 to 1,500 meters above sea level. In Beew, forests are found in 16 different locations in the northern, northeastern, eastern and southeastern parts of the village. They tend to be dominated by pine or dipterocarp trees, although sometimes combinations are found.

In 1990 Tubo registered a population of 4,589, distributed among 829 households. The majority of inhabitants are from the Tingguian linguistic group, and members of the Maeng and Kankanaey linguistic groups are a minority. The study sites are inhabited by the Maeng tribe who, like the Sagadans, are governed through the *dap-ay*. Local people farm, fish, hunt and gather forest products. A general state of poverty is aggravated by the lack of basic social services such as roads, health and educational facilities, and income-generating opportunities. The respondents are from Beew *sitio* (a 'settlement' or subdivision of a *barangay*) in Alangtin *barangay* and Bana *sitio* in Kili *barangay*. Both *sitios* are relatively new settlements; Beew was founded in 1943 and Bana in 1990. Both settlement populations had migrated into the area following disasters, Bana after the 1990 earthquake and Beew after a measles epidemic. These areas can only be reached on foot.

Of the five areas, the ecosystem of Mount Sisipitan is the most diverse, followed by Bana and Beew. Even as a secondary forest Bana retains its bio-diversity because of the minimal impact of people in the new settlement, but portions of the Beew forest are being reduced to grassland by intensified *kaingin* (swidden) farming. Because they are dominated by planted pines, Bugang and Demang forests have low diversity levels. Nevertheless, numerous plant and animal species that are of importance to local people were identified within the five areas of research: 33 timber and fuel species; nine water-bearing plants; 36 medicinal plants; 20 game animals; and 15 mushrooms. Documentation of edible fruits and species used for fibre, pesticides and honey production is still in progress.

Ownership and Land Use

The forests of Mount Sisipitan, Bana and Beew are communally owned, shared by villages and tribes – five subgroups of the Kankanaeys and two sub-groups of the Tingguians – living in the foothills of these mountains. The residents of Beew and Bana figured prominently in the protests against the transnational logging company Cellophill Resources Corporation during the 1970s and 1980s. Demang and Bugang forests, on the other hand, are parcelled into family and clan properties (Table 4.1)

Table 4.1   System of ownership in the five study areas

Site	System of forest management		
	Communal	Clan	Private
Beew	X		
Bana	X		
Mount Sisipitan	X		
Demang		X	X
Bugang		X	X

While the forests are mainly used for subsistence purposes, cultural and religious aspects are also important. Table 4.2 shows uses of the forest in each of the areas covered by the study, while Table 4.3 applies the perspectives of gender and age. All communities depend on the forest for their fuel needs, although Demang derives only a third of its fuel from the forest as residents can afford the cost of liquified petroleum gas. People in Bana and Beew rely more on herbal medicines, while the residents of Bugang and Demang, who have easier access to hospitals and other health facilities, rely on Western medicines. Women participate in *uma* or *kaingin* farming, assisting in the hauling of lumber, and gather food and medicinal herbs.

Table 4.2   Uses of the forest

Uses of the forest	Sisipitan	Beew	Bana	Demang	Bugang
Source of fuel	X	X	X	X	X
Source of medicinal plants	X	X	X	X	X
Pasture land	X	X	X	X	
Hunting grounds	X	X	X	X	X
Shifting cultivation	X	X	X		
Source of timber	X	X	X	X	X
Watershed/watersource	X	X	X	X	X
Tourism				X	X
Sacred tree site				X	X
Honey gathering		X	X		
Source of food (seasonal)	X	X	X	X	X
Agricultural land expansion				X	
Source of rattan	X	X	X		

Table 4.3   Forest utilisation by gender and age

Activity	Men	Women	Children
Fuel gathering	X		X
Hunting	X		X
Honey gathering	X		X
Uma farming	X	X	X
Performance of ritual	X		
Lumbering	X	Help in hauling	
Food gathering	X	X	X
Herb collection	X	X	

Indigenous Forest Management Systems

The indigenous concept of forest ownership recognises three tenurial arrangements. In the *saguday* or private system, forests are privatised if one has made permanent improvements such as planting regular crops or building stone walls. Owners might be members of a nuclear family who have the responsibility of protecting and maintaining the forest. Similarly, clan ownership is acquired through labour invested by the collective efforts of clan members. For the day-to-day management of the forest, a caretaker, usually a male clan member, is appointed by consensus. He has the responsibilities of mobilising clan members for tree-planting activities, deciding what types and quantities of timber can be cut by clan members at what times, and convening clan meetings to discuss matters pertaining to forest management. With such responsibilities, the caretaker receives privileges such as a larger share in forest products and priority access to resources – but he is not permitted to sell these. Both the privately owned and clan-owned forests are shared with community residents to the extent that people may gather food and branches for fuel and, with permission of the owner or caretaker, cut three to five trees for lumber free of charge.

A third system of ownership is communal, with elders deliberating on policies that are approved by consensus. In communally owned forests, watersheds are off-limits for farming and logging. Residents may only fell trees to build houses within the *ili*, or village area, or on special occasions such as the performance of rituals, weddings and funerals. Hunters must share the meat of four-legged animals with villagers who, in turn, are responsible for the hunter in case of an accident. (Traditionally, hunters used *bitu*, or pit traps, and most of the meat was shared with the community. These days, rifles and guns are used and there is less sharing, with portions of meat occasionally set aside for sale.) Forest burning is prohibited because it has been observed that rats, insects, and other agricultural pests increased whenever forests were burned. The damage caused was often significant enough to cause a food shortage. Moreover, as houses are constructed with wood and *cogon* grass (*Imperata cylindrica*), forest fires deprived people of building materials.

Whether *saguday*, clan-owned or communal, forests in Sagada and Tubo are used for a multitude of purposes. Village-based, small-scale logging is usually governed by the rule of selective cutting and limited quantity. This method has not yet reached a destructive stage, with axes and chainsaws being the main tools. Ironically, this indigenous mode of forest use is considered illegal by government agencies like the Department of Environment and Natural Resources. Another common use of the forest in Bana and Beew is seasonal honey gathering to supplement income. Honey gatherers smoke out the bees from their hives and then harvest the honey and other by-products. They have specialised knowledge of plants that emit smoke with no deleterious effects on the bees. As a measure to promote sustainability, flowering trees which are vital for honey production are protected and the use of petrochemicals on agricultural crops is discouraged or even prohibited. Although forest burning has been banned by communities in some areas, the practice continues as a way of creating grasslands for cattle grazing. This can kill or hamper the growth of trees, and can be even more destructive if the fire spreads. As a preventive measure, there is a need to develop an alternative technology for pastureland development. Finally, *kaingin* farming is a traditional agricultural system still widely practised by Bana and Beew residents. While *kaingin* may have been appropriate in the past, the pressure of the cash economy renders it inappropriate today.

The system of knowledge and technologies in the Cordillera, like those of other indigenous peoples around the world, are interwoven with beliefs and practices that are frequently dismissed as unscientific. The continued observance of these beliefs and practices, however – despite the intrusion of Western culture through religion, education, and media – distinguishes the Cordillera people from the rest of the Filipinos. Practices that may seem irrational are sometimes easily explicable in ecological terms. For example, it is not permitted to pasture in the vicinity of a spring as this would displease the spirits dwelling there. This has, of course, a direct implication for maintaining the quality of drinking water. The taboos against cutting trees when one hears the croaking of a frog might reflect concern about protecting the watershed, as frogs indicate the presence of water. The common practice of sacrificing a chicken in the forest stems from the belief that forests are inhabited by spirits. Such a practice has been very effective in imparting the message that forests have to be respected and that people may not burn the forest or dump garbage near springs. Other beliefs and practices need to be studied further to uncover their underlying logic.

A Comparison of Customary and State Legal Systems

Land is a major national issue. It can be traced to the failure of the Philippine government to reform oppressive land laws enacted by the colonisers, and their perpetuation down to the present day. Taken together, these laws and decrees serve to entrench the state's ownership of most forest land and its power to exploit forest resources, at the same time diminishing the usufruct rights of those who have managed the forests since before the creation of the Philippine state. Inhabitants of the Cordillera are now perceived as squatters on the land which their ancestors have occupied, defended and nurtured from time immemorial.

The only land rights recognised by the national government are those legally sanctioned by the state with documentation signifying private ownership. Any land that lacks this title automatically belongs to the state. The problem is intractable because indigenous people have not acquired titles or other proof of ownership. Traditionally, the labour they invested in the land and the recognition of their neighbours was sufficient proof. The land, forests, rivers and other natural resources were looked upon as being owned in common by the tribe or by the indigenous inhabitants. The people practised a form of communal land stewardship, viewing themselves as stewards or caretakers of the land, which was considered free to anyone who was willing to develop or till it. The indigenous system explicitly discourages the privatisation of forest lands to ensure that the majority retain access to forest resources. When permission is granted to a private individual or company to use the forest for personal profit, people's access is limited and the forest is usually degraded. But the government treats the Cordillera as its own resource base, awarding concessions and permits to outsiders who exploit the natural resources with no concern for the inhabitants.

Table 4.4 A comparison of official and customary views of land tenure

Issues	Government system/ State laws	Indigenous system/ Customary laws
Ownership/tenure	• Forests are owned by the government (Public Land Act, PD 705, others). • Recognises and encourages private ownership, allows an individual to own large tracts (depriving others of the resources). • Can be transferred easily through selling.	• Recognises private, clan and communal ownership. • Discourages private forest ownership, access of people to resources is open. • Land is not considered a commodity.
Tenurial recognition	• Only land titles are recognised as proof of land ownership.	• Shared knowledge of elders and community.
Acquisition of land	• Through land titling.	• Labour investment and inheritance.
Formulation of policies/laws	• Imposed from national policies.	• Discussed by elders and approved through consensus.
Process of solving land disputes	• Court procedures where money is required.	• *Dap-ay* system where money is not needed. • Consultation/collective investigation.

Meanwhile, the Department of Environment and Natural Resources, whose mandate it is to protect and enhance the quality of the country's environment, has numerous well-financed programmes for forest development. But none of these has been implemented in the five areas studied by the Montanosa Centre. Though communities are badly in need of financial and technical assistance, their willingness to be involved in these programmes is certain to be construed as recognition that the land is owned by the state. This would imply that the state can decide how land is to be used even when this goes against the people's wishes, as when the government has favoured mining companies and dam projects in the past.

Traditional and official mechanisms of conflict resolution are equally at odds. Among the Maeng and Pidlisan tribes, conflicts over tribal boundaries are resolved through the *bodong* system, or peace pact, that governs relations between the two tribes. Land disputes are resolved through a series of discussions and negotiations until a decision favourable to all is achieved. A similar process is used in Sagada, although negotiations are conducted through the *dap-ay* system. Conflicts between individuals, including disputes over inheritance, are effectively resolved through the *dap-ay*. By contrast, the current legal system resolves conflicts through the courts. This system favours those who possess documents, sometimes obtained by deceitful means, and can afford the best legal counsel.

For several generations now, the forests of Beew, Bana and Sisipitan have provided for the needs of the people. In turn, the communities have developed a system of conservation and management. In other words, the forests of Demang and Bugang are the concrete manifestations of people's efforts to improve their environment. While at present they may not be replanting trees actively, they claim that natural regrowth takes place as long as the forests are protected from fire. For decades, the interplay of the indigenous socio-political systems and the values and wisdom of local people resulted in the protection of forests that are today considered to be among the last frontiers of the country. Unfortunately, modern developments have taken their toll on people and the forests. Several factors are contributing to the erosion of indigenous forest management systems.

State forest laws continue to undermine indigenous concepts of natural resources management. As local systems and leadership are weakened, the honesty, justice and culture of sharing also starts to disintegrate. Local people find it hard to cope with the intrusion of the cash economy into the traditional subsistence economy. There is also a lack of developmental opportunities. To meet their economic needs, the people of Beew and Bana resort to intensified *kaingin* farming and hunting, which increases pressure on their forests. In Sagada, the shift from subsistence agriculture to commercial vegetable and orchard production has led to the conversion of pine forest into agricultural lands. Moreover, an increasing number of small-scale loggers, equipped with chainsaws, cut and sell timber in violation of selective cutting rules. Pressure on forest is intensified by the government's failure to provide basic social services. In Beew and Bana, for example, there are areas with a high potential for more efficient agriculture. Unfortunately, the area lacks adequate irrigation facilities and people are forced to continue with their *kaingin* farming.

Local Initiatives

Growing awareness of the contradictions between state policies and indigenous management practices has led, on the regional level, to a revival of people's organisations, most of which are members of the Cordillera Peoples Alliance for the Defence of Land and Resources and for Self-determination. The Alliance is guided by the belief that ancestral land is fundamental for indigenous peoples. Land is life. The land and people are one – a collective and integrated whole. The land and the people comprise the *ili* or village. Boundaries are upheld not only by the community but also by the adjoining and even distant villages and tribes. Indigenous peoples uphold the principle of sharing and nurturing nature's bounty.

In 1993 more than 90 representatives from the different *barangays* of Sagada convened in a forum to discuss three government programmes: the Certificate of Ancestral Domain Claims, the Certificate of Ancestral Land Claims, and the Certificate of Land Ownership Awards. The people ended the forum by rejecting all three programmes and presenting a petition demanding that the government respect indigenous laws. Subsequently, some officials of the Department of Environment and Natural Resources compromised by allowing people to cut timber in their forest on condition that the timber was used within the *ili*. This was a small but significant gain in the struggle for recognition of customary law.

From the mid-1970s to the mid-1980s, people in Bana and Beew were at the forefront of the struggle against the Cellophill Resource Corporation, a transnational logging concern favoured by the Marcos dictatorship. Many lives were sacrificed to protect the forests for succeeding generations. Had it not been for the relentless efforts of the Tinggians, Kankanaeys, Bontoks and Kalingas, Cellophill would have wreaked ecological havoc on the forests of the Cordillera. Even then, significant damage was inflicted before the company stopped its operations. Since then, local people have started an annual collection of rattan seeds for planting in the month of August. They have also embarked on a reforestation project and a search for alternative farming technologies suitable for sloping areas.

In Pidlisan the *dap-ay* retains a strong influence, and a people's organisation, the Asosasyon dagiti Sosyudad ti Umili ti Pidlisan (ASUP), was formally launched in 1990. Included in the organisation's nine-point programme are the development of indigenous socio-political systems and the conservation, protection and development of natural resources within the ancestral domain. Earlier, in 1983, the people of Pidlisan had banned *kaingin* on Mount Sisipitan. Six years later, they formulated and executed a policy restricting small-scale mining to designated areas within their ancestral land. In 1994 they comprehensively rejected the Small-Scale Mining Act.

Recommendations

After more than 15 years of grassroots development work, the Montanosa Research and Development Centre has formulated a number of recommendations to Cordillera policy makers and development agencies engaged in forestry reform in the Philippines.

Conflicts over land and resources must be resolved. The land rights of indigenous people and their customary laws must be recognised by government. This requires a genuinely autonomous regional government that would give the peoples of the Cordillera the right to determine freely their political, economic and cultural ways of life.

Community organisation has proved effective in empowering indigenous communities and should be adopted as an inherent part of development work. This includes forming alliances or federations from village to provincial levels, and joining the regional movement for the assertion of indigenous people's rights and self-determination.

Traditional practices that serve the long-term interests of people with regard to ownership, use and management of forests should be strengthened or revitalised. Policy makers should support legal and indigenous institutional arrangements which prevent individuals or outside groups from exploiting natural wealth at the expense of the poorer sections of the community. They should recognise the role of the people in the development and management of the natural resources on which they rely for many of their basic needs. Cooperative efforts in the utilisation and management of the forest resources should be encouraged. This requires the definition of objective parameters for a qualitative assessment of forested areas that measures the extent of forest cover, its quality and dynamics, and its stage of regeneration or degeneration. On the basis of this assessment, communities could then develop appropriate forest-use and development plans.

BIBLIOGRAPHY

Community Health Education, Services and Training in the Cordillera Region (1989) *Common Medical Plants of the Cordillera*, Baguio City, Philippines.

Cordillera Peoples' Alliance (1995) 'Ancestral Land Delineation and Indigenous Peoples', unpublished position paper.

De Padua, L., Lugold, G. and Pancho, J. (1977–83) *Handbook on Philippine Medical Plants*, Vols

1 (1977), 2 (1978), 3 (1982) and 4 (1983), University of the Philippines-Los Banos, Laguna.

Gonzales, Pedro and Rees, Colin P. (1988) *Birds of the Philippines*, Haribon Foundation for the Conservation of Natural Resources, Inc.

La Vina, A. (1991) *Law and Ecology*, Legal Rights and Natural Resource Centre, Manila.

Merrill, Elmer D. (1912) *A Flora of Manila*, Department of the Interior, Bureau of Science.

Merrill, Elmer D. (1926) *An Enumeration of Philippine Flowering Plants*, 4 vols, Manila.

Montanosa Research and Development Centre (1993) 'Preliminary Community Appraisal of Beew', unpublished report.

Municipal Planning and Development Coordinator (1995) 'Sagada Municipal Profile', unpublished report.

Quimio, T. (1978) *Common Edible Mushrooms in the Philippines*, University of the Philippines-Los Banos, Laguna.

Quisumbing, E. (1978) *Medical Plants of the Philippines*, Quezon City, Philippines.

Rabor, D. (1977) *Philippine Birds and Mammals*, University of the Philippines Science Education Centre, University of the Philippines. Press, Quezon City.

Tauli, A. (1984) *Dakami ya Nan Daga Mi*, Baguio City, Philippines.

5 Alternatives to Rainforest Logging in a Chachi Community in Ecuador

LORENA GAMBOA

Acción Ecológia, Ecuador

Geography, Ecology and People

The Centro El Encanto Reserve, inhabited by both indigenous Chachi and Afro-Ecuadorian people, is located along the Cayapas River border in the province of Esmaraldas, northwestern Ecuador. The area contains the last tropical rainforests on the Ecuadorian coast, forming part of the bio-geographical region of the Choco, which has one of the world's highest levels of biodiversity and endemic species. Centro El Encanto is part of the Cotacachi-Cayapas Ecological Reserve buffer zone. It is estimated to harbour a total of 6,500 plant species, equivalent to 25 per cent of all plant species recorded in the country. Of these plants, it is estimated that some 1,260 species are endemic to the area.

The native vegetation of Cayapas has the appearance of a dense ever-green forest with species of great height and width. Forest composition varies according to the condition of the soils, drainage, topography and humidity. The most important of the tree families are *Fabaceae, Moraceae, Lauraceae, Myristicaceae*, and *Meliaceae*. The climate of the Cayapas River area is hot and humid. The average temperature is approximately 25°C and the average annual rainfall 4,000 millimetres. The area is characterised by an extensive network of rivers and streams, which also constitutes the principal system of communication, transportation and trade for local people.

For more than 400 years two distinct ethnic groups, the Chachi and the Afro-Ecuadorians, have occupied the forests of the region. Over this period, both the Chachi and the Afro-Ecuadorians, through their respective cultural practices, have managed the forest sustainably, providing themselves with food, clothing, medicine and ritual necessities. Both groups also practise small-scale agriculture and have developed an in-depth knowledge of the use of forest plants and the hunting of wild animals. The horticulture practised by the Chachi is well adapted to the conditions of the tropical rainforest, requiring no external inputs. Their *fincas* or *chacras* (fields) are rich in a variety of products, and support a wide variety of animals and extensive ground and soil cover. In general, the two groups have lived together in peace, with the exception of sporadic conflicts rooted in cultural differences. In the last few years, however, the presence of outside agents in the area has intensified conflicts by promoting competition over resources, especially timber. The utilisation of the forest based on traditional knowledge and cultural practices is changing rapidly. The process of acculturation has led to increased logging and destruction of the ecosystem. Nowadays, the mainstays of the Chachi way of life are agriculture and the extraction of wood. Hunting, fishing and collection of forest products have been relegated to secondary activities.

There are several pressures on the forests and other natural resources. As these forests are denser than the forests of the Ecuadorian Amazon, the region has become well known for its potential as a rich source of timber. This has resulted in the presence of a large number of logging companies, all looking for high-quality timber to supply the national and international markets. As unprocessed logs fetch very low prices, there is an incentive to fell large areas in order to increase revenues. The logging companies encourage local Chachi and Afro-Ecuadorian communities to participate in the extraction of wood. The government promotes the exploitation of the forests for export, but lacks the personnel and resources to regulate such exploitation or to force the companies to abide by their management plans.

Colonisation is encouraged by the state and facilitated by logging companies that open the forest with their access roads. The expansion of agro-industry, in particular monocultural plantations of banana, *ariucan* oil palm and *palmito* (palm heart), has also replaced vast areas of natural forests. The development of road transportation adds to the influx of entrepreneurs who seek to exploit the wealth of the forests. Insecurity of land tenure, combined with the growth of both the indigenous and colonist populations, causes increased pressure on the forests and provokes conflicts among the various groups. Meanwhile, state forest policies jeopardise the overall management and conservation of the remaining native forests. The opening up of markets for logs and the plan to lease national forests to private logging companies is promoting a higher level of exploitation and imposing a market-based model of development. Forests are being seen exclusively as sources of timber.

Within this context, the Centro El Encanto Reserve has become a reference point for other Chachi communities because of its firm decision not to lease out its forests and to opt for methods of development more closely linked to the communities' cultural values. These include, for example, the production and marketing of handicrafts, community management of ecotourism, and the sustainable extraction of non-timber forest products (NTFPs). The Chachi in Centro El Encanto still depend for up to 85 per cent of their diet on their own crops. They derive 90 per cent of their supply of animal protein from hunting small and medium-sized rodents that feed on the products grown in their *chacras*. It is crucial to maintain and improve the management of these practices, and to find ways of adding value to the products being grown.

The Chachi have several ways of managing and using forests which are suitable to the cultural and ecological conditions and can be developed further. For example, a number of NTFPs can be identified, including rare and potentially very valuable medicines. The knowledge and use of medicinal plants and rituals is the special field of the shamans (*mirucus*), while the rest of the population knows little of the properties or potential use of these plants. An increased focus on forest products, such as those used by the *mirucus*, could be an instrument for recovering cultural values linked to the knowledge of the forest. Meanwhile, several better-known NTFPs already have potential on the national market. But costs of transportation are high, and non-Chachi middlemen control the trade. The creation of a system of community marketing with transportation and access to regional and national markets is therefore an urgent need. Only when these conditions are met will the Chachi receive a fair price for their products and experience the economic benefits of managing the forest for NTFPs. In order to reduce the incentive to log, meanwhile, it is necessary to improve community timber marketing structures for those communities that have already become part of the forest exploitation dynamic. If the community receives fair and equitable prices, the level of exploitation can be reduced.

Community members are engaging in a process to elaborate a forest management plan and to develop a set of rules and sanctions to control access to communal areas. We envisage that the increasing pressure on these forests – and the conflicts among community members that result – can be diminished.

Culture and Social Organisation

The Chachi are one of the original peoples of Ecuador. Their population is about 5,500. They are considered to be a peaceful people: throughout their history they have preferred to migrate rather than become involved in conflicts. They are identified as a culturally distinct ethnic group that maintains a lifestyle based on the knowledge and use of the forest. Many traditions and customs regulate their behaviour and world view.

The social organisation of the Chachi is based on the nuclear family. Chachi families consist of husband, wife, children and dependants. All members of the family have duties and obligations which vary according to age and gender. The woman, however, carries the heaviest burden of work. Women are responsible for a great variety of household tasks, such as the collection of firewood and the tending of food crops (work which is shared with the children and the in-laws), the preparation of food, making and washing clothes, the cleaning of the house, work in the orchards, gathering of products from the forest, fishing for river shrimps and clams, care of domestic animals, and the making of handicrafts such as baskets, mats and fans. Last but not least, they raise the children and guard over the health of family members. The role of women in conservation and the transfer of knowledge to the next generation cannot be overestimated. They continue to practise traditional activities and tend to be sensitive to the need to maintain the forest's integrity. Chachi men, on the other hand, have lessened their involvement in traditional activities and spend increasing amounts of time cutting the forest. The gap is thus growing between traditional concepts of the forest within the community and the values associated with the social and environmental changes brought by commercial forest exploitation.

Two distinct forms of social organisation currently guide the life of the Chachi – the traditional and the politico-legal. Though interrelated, the two systems influence Chachi society and the natural environment in different ways. Within the traditional form of organisation there are two key players: the *uni*, or cultural leader of the Chachi, and the *chaitalas*, the representative of the *uni* in the community. The *chaitalas* is charged with supervising the observance of traditional laws, especially regarding ethics and respect for resolutions taken in the public assemblies. It is the *uni's* obligation to maintain traditional law, which acts as a mechanism for social structure, integration and cohesion. Adherence to traditional law plays an important role in the maintenance of Chachi culture and the recognition and respect for community authorities (*unis* and *chaitalas*) and shamans (*mirucus*).

The politico-legal organisation maintains and strengthens relations among communities, and governs their dealings with the outside world. The principal authorities elected in the popular assembly are the governor and the community council, which consists of a president, vice-president, treasurer, members and substitutes. The main functions of the council are to represent the community with state authorities, NGOs and churches.

The Chachi are grouped in 27 centres, some of which are composed of several communities. These form the Federation of Chachi Centres of Esmeraldas (FECCHE), which is affiliated to the Confederation of Indigenous Nationalities of Ecuador (CONAIE). The mother tongue of the Chachi is Chapalachi, which is still the main language of communication within the community, though most people also speak Spanish.

Ownership and Forest Management

Between 1968 and 1992 the Ecuadorian government granted land rights to settlers and indigenous people in the tropical northwest region. Among those receiving land was the indigenous community of the Chachi. Currently, the Chachi territory encompasses an area of approximately 89,501 hectares where some 10 communities have legal title to their land. Within each centre land has been distributed in accordance with the size of the area and the number of families that make up the community. With an area of 8,000 hectares, Centro El Encanto has one of the largest allocations of land.

El Encanto is made up of three communities: Encanto, Rampidal and Santa Maria. The land is demarcated and people have property titles. Of the total 8,000 hectares, 2,000 hectares form a community reserve. This is surrounded by some 120 family holdings of approximately 60 hectares each. Each family consists of about five to six people. The ownership of the central community reserve rests with the entire community, and ownership of family holdings with each family.

In the case of the Afro-Ecuadorian community, land allocation has been much more haphazard and has contributed to conflicts in some areas. In the Centro El Encanto, 30 Afro-Ecuadorian families have been given a total of 679 hectares by the Chachi on the basis of an agreement between the two communities. This land, however, has not been distributed equally among members of the Afro-Ecuadorian population.

The community forest reserve has not yet been subject to logging, agricultural clearings or other forms of encroachment, but the 60-hectare family plots are subjected to two forms of much more intensive use, mostly on the basis of traditional land-use patterns. First, the *chacras* are 8-hectare patches of land in partially cleared forest which are cultivated intensively. Usually they consist of tree crops, vegetable gardens or open land, and tend to produce about 80 per cent of the family's animal and vegetable food (see Appendix 2). The major crops in the *chacras* are cacao (*Teobroma bicolor*) and coffee (*Coffea arabica*), to which 1–5 hectares are devoted and which are traded in the local market. Plantain (*Musa sapientum*) and yuca (*Manihoc sculenta*) are grown for self-consumption, though a portion of the plantain crop is marketed.

Second, the canopied forest, used less intensively, yields construction timber, wild fruits, fibres, medicines and some animal protein (see Appendix 2). Products from the forest include timber, which represents the most important source of monetary income for the Chachi, and handicrafts made from forest fibres (baskets, mats, cloth and fans).

Apart from chickens, and pigs in a few cases, the Chachi do not practise much animal husbandry. The principal source of animal protein is game, which is said to have better flavour than domestic animals. For hunting, the most productive areas are the *chacras*. This is probably because the *chracras* resemble an early stage of forest regeneration and are therefore favoured by various species of rodents.

Timber Extraction

Traditionally timber extraction has been limited to local use. Spurred by the inroads of the timber industry into the area, however, logging is increasing rapidly in the family holdings and eventually may affect the community forest reserves. Most of the Centro El Encanto community reserve has remained untouched, protected by its distance from the river, which is the main avenue for the transport of logs. Of late, however, there have been intrusions by the Afro-Ecuadorian community who consider the reserve to be 'unmanaged land'. The reserve is also under increasing pressure from a few Chachi families who have been persuaded by the timber industry to supply logs. If no measures are taken, a confrontation between the Afro-Ecuadorian and Chachi communities, or among the Chachi, is likely to occur.

Meanwhile, the main external pressure on El Encanto forest comes from the timber industry, which sends agents to the community to offer individual people rewards for logs delivered to their collecting yards. The logs are cut either by landowners – or, in the case of the central reserve, mainly by Afro-Ecuadorians – and floated downriver to the collecting yard. A low price is paid for logs and an accordingly large area of forest must be felled to provide a reasonable income to landowners or labourers. Another method used to obtain community timber is the offer of community amenities – a road, a house or a school – in exchange for timber sold at a price set by the industry. This type of timber extraction is destructive, especially as no regeneration or replanting activities follow the cutting.

Another pressure on these forest ecosystems is the influx of settlers into the area, facilitated by the logging roads. The settlers are mostly *mestizos* from Manabi and Loja provinces. They claim land by completely clearing and burning the forest.

Government Policy

Instituto Equatoriano Forestal de Areas Naturales y Vida Silvestre (INEFAN), the government department charged with the management and preservation of forests and protected areas, promotes the exploitation of forests for export without the necessary infrastructure to regulate such extraction activities. The state system lacks resources and personnel to control logging, and to ensure that companies abide by management plans. Furthermore, the government does not provide incentives for conservation and reforestation, or support local production initiatives and traditional forest management. The state has a responsibility to check unplanned colonisation in order to avoid conflicts with the local population and to protect the native forests.

INEFAN has been leasing forests to private logging companies, which constitutes a serious threat to the remaining native forests. Multilateral agencies have also financed reforestation projects such as PLANOFOR and PROFAFOR. But these consist largely of subsidies (up to 75 per cent) to plant trees, and are restricted to timber species, while the need to regenerate the area's biological diversity is ignored.

It should be mentioned that the current government has approved the export of timber from 'sustainably managed' native forests. But there exists no model of sustainable forest management in the country, and this measure will only increase the demand for high-quality wood and accelerate the destruction of tropical rainforests on a large scale.

Predicted Changes and Impacts

If present trends continue, the forest managed by the Chachi communities will be degraded beyond any possible regeneration within a few years. The continued loss of timber resources may also lead to conflict between and within the various groups. As resources diminish, Chachi can be expected to defend their rights more vigorously. The continuous influx of colonists is another likely source of future conflicts.

In Chachi family holdings, the percentage of land covered by *chacras* will probably increase. The subsequent reduction of canopy cover in the traditionally managed lands will result in a reduction of ecosystem stability and natural biodiversity. As the forest disappears, the Chachi's main source of basic needs and income also vanishes. This will lower the standard of living and turn them into a poor and marginalised people. A loss of cultural values and traditions is likely to follow the loss of the forest.

Local Initiatives and the Role of Outside Parties

Although the trends are worrying, Centro El Encanto is today a model of alternative development. The community's firm resistance against logging companies and its active role in denouncing the abuses of the timber industry have given it a special status. El Encanto has the potential to encourage other communities to follow its example.

During its struggle, El Encanto received the support of various NGOs and networks. The national network of indigenous peoples' organisations, CONAIE, lent decisive political backing in the fight with the logging company ENDESA-BOTROSA, and helped to negotiate with the state.

Together with environmental organisations, CONAIE helped to define the situation of the Chachi centres and to assess their principal needs. El Encanto was chosen to implement a pilot project of raising small domestic animals, such as chickens and pigs.

The NGOs collaborating most closely with Centro El Encanto are Fundacion para el Desarrollo Alternativa (FUNDEAL), Fondo Ecuatoriano Populorum Progressio (FEPP) and Acción Ecológica. (In addition, the Camboniana mission has a base in the centre, which for a century has focused on education alongside religion.) All these NGOs have backed the inhabitants of El Encanto against the destruction of the forests and the leasing out of lands.

As the Chachi communities look to the future, the campaign against deforestation and the leasing out of land for logging will be strengthened and broadened. To that end, the leadership of FECCHE, the federation of Chachi communities, will be strengthened to make the organisation more independent from the interests of the logging companies. Furthermore, an integrated community forest management plan is being developed. Community rules to regulate access to the forest will be drafted. There are a number of projects to enhance community self-reliance in the areas of food production, the management of the *chacras*, the promotion of the quantity of available game, the production of marketable products, aquaculture and ecotourism. There are also plans to develop a system of community marketing, transportation, local and national shops for the sale of Chachi handicrafts, training to improve the quality of handicrafts, and research into NTFPs with economic potential. Finally, there are initiatives to promote Chachi culture and traditions through measures such as the strengthening of bilingual education.

6 Protecting the Last Primeval Forests of Paraguay

The Cordillera del Yvytyrusu and the Survival of its People

SOBREVIVIENCIA

Paraguay

The humid subtropical climate, geographical isolation, and fertile soils of Paraguay's Cordillera del Yvytyrusu have resulted in an ecosystem rich in species diversity and endemism. Sobrevivencia's research suggests, in fact, that these forest areas represent a relic from a pre-glacial era. They also contain a number of archaeological sites with pre-Columbian inscriptions that have yet to be deciphered. Yvytyrusu is also the traditional territory of the Ache-Guayaki and the Mbya-Guarani peoples, who have for decades faced violent persecution and forcible displacement by the government, peasants and large landowners.

But these forests are being systematically destroyed and replaced by cash crop agriculture and pastures for cattle grazing. Given the area's unique characteristics and fragile conditions, Sobrevivencia has proposed that designation of this mountain range as a biosphere reserve would afford the best protection to the area's outstanding biological and cultural values. This conservation measure would be conditional on granting the surviving Ache communities, who were expelled from the area, the right to reclaim land in Yvytyrusu adjacent to the proposed core zones.

Unique Geography, Fragile Ecology

The Cordillera del Yvytyrusu mountain range system is located at the centre of eastern Paraguay in the provinces of Guaíra and Caazapá. It includes the country's highest peaks, around 840 metres above sea level, and forms part of the watershed between two of the largest river basins in South America, those of the Paraguay and Paran·rivers. Numerous creeks and small rivers that find their sources here depend on the survival of the mountain forest cover. The climate in the area is subtropically humid, with warm, very humid summers and cool, relatively dry winters. The mean annual precipitation is about 1,600 millimetres, with the most humid period in October–May.

Within a massive sandstone base from the Older Jurassic period, there are thick layers and chimneys of lava dating from the Cretaceous. Basalt caps protected the underlying sandstone from erosion for millions of years, so that the existing range is sculpted out of the surrounding landscape. The numerous creeks have carved deep ravines in the sandstone, forming waterfalls at the end of the basalt layers. Weathering of this basalt produced very fertile soils. But because of the slow weathering process and the steepness of the slopes, the soil layers are very thin and prone to erosion. In some parts, sandstone has been modified by lava heat into hexagonal columns.

Entomological research, carried out in the upper reaches of the Tacuara Creek basin and on Cerro Acatï in the largest and least disturbed forest of the northern ranges, demonstrates that these mountains hold an extraordinary diversity of life. Indeed, many of the species found in the area by Sobrevivencia and other specialist researchers are new to science. From an ecological point of view, the Yvytyrusu range, together with the northern border region adjacent to Mato Grosso in Brazil, is probably the most important region in the whole country.

Results of Entomological Research

Two families of nocturnal butterflies – *Sphingidae* and *Saturniidae,* both the subject of recent important monographs resulting from the entomological research carried out by Sobrevivencia – will serve as examples. Species of the *Saturniidae* family have been found here, for which the previously known distribution was limited to the Serra do Mar, the Atlantic coastal range in southern Brazil, and the lower altitudes of the Andes in Bolivia and northern Argentina. At the Serra do Mar and on the eastern Andean slopes, part of the original vegetation survived during the dry periods when the savannahs dominated in South America. These dry periods coincided with the glaciations in the northern hemisphere during the Late Pliocene and Early Pleistocene periods. The reinvasion of Paraguay by the subtropical humid Atlantic forest is relatively recent. It began between 10,000 and 12,000 BC. The detection of these species of butterflies in the Cordillera del Yvytyrusu suggests that native flora and fauna survived during the glacial periods. This would explain their notoriously disjunct distribution. They probably represent relics that survived in local refuges where the climate remained favourable during extreme climatic changes in the wider region.[1]

Research on the *Sphingidae* fauna in Paraguay showed the existence of 83 species, of which 72 were found in Yvytyrusu. With this quantity, Yvytyrusu has the greatest diversity in the whole country. The studies on *Saturniidae* show similar tendencies. Of the *Rothschildia* genus, six of the seven known species in southeastern South America live in the Cordillera del Yvytyrusu. The discovery of these butterfly species proves that the

Yvytyrusu ecosystem is a part of the original vegetation that survived during drier epochs when the savannahs predominated. Several animal species also survived and these are endemic today. Furthermore, with the advance of the Atlantic forest from the east during the last post-glacial epoch, which had its western limits at the Cordillera del Yvytyrusu, the local ecosystem was further enriched with additional species. When the climate turned more humid, the rocky formations on the slopes and mountaintops became a refuge for many xerophytic plant species which were typical of the dry periods (for example, there are several species of the cactus family). A rich ecological system with unique species diversity was thus formed.

As these species are found solely within a few isolated geographical zones, it is evident that the area possesses the ecological conditions required for their survival. These zones, small in size and dispersed, must once have been vast and contiguous with a continuous distribution of species – since continuous genetic exchange within the population of one non-migratory species is only possible if that species has continuous geographical distribution. The period during which such fluid exchange took place is relatively recent because today's isolated populations of these species have developed only negligible differences in morphological characteristics. In fact, it appears that the Cordillera del Yvytyrusu acts as a refuge and the endemic species found there have the character of relics. This is one reason why the protection of the Cordillera del Yvytyrusu should be made an immediate priority. Since the survival of this system depends directly on the fragile stability of its particular climate, which is being maintained by its own vegetation, strict conservation measures are needed in the whole southern massif, the central massif (Cerro Acatï) as well as in all the mountaintops, steep slopes, upper watersheds and water-courses.

In addition to its ecological importance, parts of the Cordillera del Yvytyrusu should also be strictly preserved for archaeological reasons. One area that deserves special attention is the Cerro Polilla at the outlet of Tororö Valley where as yet undeciphered cave inscriptions were registered recently by Sobrevivencia. Peasants living in the north of the range introduced us to a system of caves in the Takuara Creek Valley decorated with pre-Columbian inscriptions that are probably related to those in Tororö Valley. This latter site is located between two peasant communities, Tercera Fracción and Mainumby, some 10 kilometres from the town of Melgarejo. These archaeological sites represent a fundamental contribution to the largely unknown prehistory of the Paraguayan territory.

Moreover, according to the new constitution, the last group of Ache people who were expelled from the forests of Yvytyrusu (Ache Purä) have the right to reclaim their ancestral lands. These people are today divided into several communities north of the Yvytyrusu – in Cerro Morotï, San

Joaquín, Caaguazu Province and Chupa Pow community in Canindeyu Province near the northeastern border with Brazil. Any management plan for the Cordillera del Yvytyrusu should respect and incorporate this right.

Taking into account all these conditions, we propose that a designation of the Cordillera del Yvytyrusu as a biosphere reserve is the best guarantee for its adequate protection. To that end, the mountain range can be subdivided into several categories within the definition of a biosphere reserve. The whole of the southern massif, which includes the highest mountain peaks and the largest tracts of forest, can be designated as core zone areas. These would cover approximately 100 square kilometres. The central massif, which covers an additional 60 square kilometres and includes the Cerro Acatï, can also be included in this category. The rest of the range, which includes areas occupied by peasant families and the mountain tops and steep slopes of the northern massif, could be classified in other categories.

The Natives and Peasants of Yvytyrusu

Until the 1970s only two indigenous peoples inhabited the forests of Yvytyrusu, namely the Mbya-Guarani and the Ache. The Mbya people are semi-sedentary agriculturalists who began to inhabit the forests and grasslands of the mountain range approximately 800 years ago. The Ache are predominantly hunter-gatherers who were already living in the forests of the inner valleys when the Mbya arrived.

Between 1960 and 1980 the greater part of both the Ache and the Mbya populations were systematically exterminated, and the survivors were displaced from their traditional territories. Currently, there are only six Mbya Guarani communities, each one composed of 20 to 30 families, inhabiting small forest patches around the range: Ovenia Santa Teresita near Paso Jovai; Nurundiary, located in Colonia Sudetia, east of the range; a small group in 4 de Diciembre near Fassardi to the south of the range; Ka-amindy, Kapi-i, Yryvuku-a and Isla Hu near Colonia Independencia northeast of the range.

The Ache indigenous communities, millenarian inhabitants of these forest-covered mountains, were persecuted, decimated and finally expelled from their ancient home. Small groups of Mbya eagerly participated in this 'hunt for savages'. Even some baptised Ache searched for their errant relatives in the area. It was always evident that the Ache had little resistance to change. Accustomed to living deep in their forests, life in open territory was very harmful to them: they simply got ill and died. In 1978 the last groups of original inhabitants of the range were forcibly evicted and moved to reserves managed by Christian missionaries. The few who survived the illnesses (especially influenza and smallpox) brought to them by their Paraguayan captors, seek out a living on small plots of land insufficient for the conservation of their traditional ways of life. Survivors of other Ache groups from the Alto Paraná Forest are also settled there. Despite the horror and suffering, the Ache from Yvytyrusu, or Ache Purä as they call themselves, maintain a strong memory and a great nostalgia for their ancestral territory, although these are mixed with memories of violence and extermination.

Many of the last Ache groups were moved from Yvytyrusu to the Chupa Pow settlement, located some 200 kilometres to the northeast, in Canindeyu Province. In their original territories of Yvytyrusu, the Ache people had about 50,000 hectares of forest connected with the great Alto Paraná Forest, which once covered over 1.5 million square kilometres from the humid savannahs along the Paraguay River to the Atlantic Ocean in Brazil. Now these groups share a 7,000-hectare tract of primary and secondary forest with Ache groups brought from other areas of the Alto Paraná Forest. The majority of the Ache Purä were taken to a community called Cerro Morotï in Caaguazu Province, on a 1,358-hectare lot, where some 48 nuclear families live. Both communities are under the tutelage of Christian missionaries.

These native communities were progressively replaced in Yvytyrusu by *mestizo* peasant families and farmers of European origin. The immigration of Paraguayan peasant families to the Cordillera began in the late 1940s and intensified during the 1960s. Already during the first years of this century European farmers, most of them of German origin, settled in the forests and natural pasture lands of the Mbya east and north of the range. Later they also penetrated the forests in the mountain range and converted thousands of hectares to pastures. Some Ache individuals who were kidnapped when they were children are still living around the Yvytyrusu. Taken away from their parents (who were often murdered), they now work as servants with peasant or farmer families. Later, absentee landowners seized the remaining state-owned forest lands, burnt the forest and started to develop pastures for cattle raising.

Sobrevivencia has so far worked mainly with peasant families living in the northern part of the range. At the same time, contacts have been maintained with people from other zones within Yvytyrusu. Our objective has been to promote collaboration among the peasant communities of the whole range, such as Santa Secilia and San Gervacio to the east, and Ita Azul, Polilla and Tororö, towards the centre of the range. In 1991, the designation of the range as a national park raised concern among local people. Peasant communities in the area responded by creating the Asociación de Pobladores del Cerro Yvytyrusu to coordinate their actions.[2] This association encompasses the peasant communities shown in Table 6.1. Peasants grow both food crops and cash crops such as sugar cane and cotton. There are some small industries, like sugar cane mills, distilleries of *petit grain* (an aromatic essence obtained from bitter orange leaves), and handicrafts such as the traditional *ao po'i* embroidery.

Table 6.1 Peasant communities in the Asociación de Pobladores del Cerro Yvytyrusu

Peasant Community	Number of families
Calle Florida	30
Tercera Fracción	26
Mainumby/Rancho Cuatro	50
Zorrilla Cué	30
San Vicente	100
Tacuarita	60
Itatí	20
San Blas	20
Total	336

Because of fear of displacement due to the creation of Yvyturusu National Park, peasant leaders asked Sobrevivencia to assist them with legal matters, and to help increase the sustainability of their agriculture. Some families have indicated that they are prepared to replace sugar cane and cotton with a more ecologically appropriate mixture of crops. Sobrevivencia is now working with the Asociación de Pobladores del Cerro Yvytyrusu to shift agricultural production patterns towards self-sufficiency. The cultivation of native medicinal and ornamental plants is also being promoted. Sobrevivencia also maintains contact with Cerro León, Cerrito and Vista Alegre communities in the north part of the range, though they are not actively involved in the Asociación, which gained formal legal status in May 1994. Sobrevivencia has supported this process from the beginning.

Legal Status of the Forest

Although most peasant families in Yvytyrusu lack property titles, they do have certain rights over the land they cultivate. Most importantly, they have the first right to buy this land from the state through the Rural Welfare Institute (IBR) provided they have occupied the land for more than ten years. But according to Paraguayan law – among others, the Rural Code (No. 1248, 1931) and the Forest Law (No. 854, 1963) – forest lands on steep slopes do not qualify for agrarian reform. This applies to parts of Yvytyrusu. It has been proposed that some peasant landholdings should be vacated as they are located on steep slopes, along rivers or in vulnerable parts of a watershed. These areas need to be reforested and protected. Large landowners, meanwhile, have purchased significant tracts of land in the area from the state, in deals involving dubious interventions by the Rural Welfare Institute.

Chapter V of the 1992 National Constitution of Paraguay, entitled 'Of the Indigenous Peoples', and the Statute of Indigenous Communities (No. 904, 1981) state that the Ache from Yvytyrusu – who now live on reservations far away from there – have the right to reclaim part of their ancestral territory. By contributing scientific, anthropological and technical information, Sobrevivencia aims to support the Ache people in their struggle to reclaim this land. One can only hope that by the time the Ache finally regain their land, some original forest will remain.

With Decree No. 5815, 17 May 1990, the government of Paraguay created the Yvytyrusu National Park encompassing 240 square kilometres of territory. The authorities have not yet demarcated the park's borders. To our great concern, however, the plan is to categorise the Yvytyrusu range as a so-called 'managed resources reserve' under the newly proposed National System of Protected Wild Areas (SINASIP). This would allow agriculture, cattle raising and forestry in the area, which undoubtedly would lead to the final destruction of the last tracts of primeval forest.

The government's latest proposal is to include the new Yvytyrusu Managed Resources Reserve in a biosphere reserve. This would also include the neighbouring – and highly endangered – national parks of Caaguazu in Caazapá Province and San Rafael in Itapuá and Caazapá provinces. These two parks have also been recategorised as managed resources reserves within the new National System of Protected Wild Areas. All three areas lie within the upper Tevycuary River watershed, the largest inner river in Paraguay.

Sobrevivencia, in collaboration with other environmental and indigenous support organisations and local indigenous and non-indigenous community organisations from Yvytyrusu, is directing a campaign to ensure the adequate protection of the whole Cordillera, with special attention to the area's ecology and traditional inhabitants. To that end, we are also scrutinising the property titles of landowners.[3]

Problems of Land Use

The forests of Yvytyrusu are disappearing rapidly. Between 1945 and 1985, forest cover was reduced by more than half. On the northern massif, only steep slopes and hilltops have forest cover, though large intact tracts of forest remain on the southern massif. Since 1985, deforestation has intensified even further. Absentee landowners are continuously clearcutting and burning, even on very steep slopes, to create pastures for cattle. Large quantities of herbicides are employed that harm both the remaining native vegetation and the small agricultural plots of the local peasants. Some of the inner valleys and slopes have recently been denuded. This process has accelerated with news that a national park will be created, and is likely to continue unless immediate measures are taken to halt it. The areas that have kept their original vegetation represent continuous forest systems that have so far endured little human influence. These are sufficiently large to allow for the survival of a unique diversity of flora and fauna.

Over several decades, peasant communities cultivated the fertile soils of the Yvytyrusu valleys. This was not particularly damaging to the ecosystem, as the population was small and land was mainly used to grow subsistence crops. The situation is changing rapidly, however. The population is growing, and people have started to consume imported goods, which forces them to grow cash crops to earn extra income. Currently, some 2,500 peasant families occupy 20 per cent of the Yvytyrusu range, which has a total area of some 300 square kilometres. Most communities are located in the north of the range, in the district of Colonia Independencia, and along a central valley that crosses the range from east to west, between Ita Azul and Tororö. Cotton and sugar cane are grown for export. As economic dependence increases, farmers are compelled to cultivate progressively larger areas, and steep, formerly forested slopes are being converted to farmland.

There have been various responses to the designation of Yvyturusu as a national park. The large landowners accelerated the process of deforestation to create pastures, under the pretext that they were adhering to the official policy of 'rational and intensive exploitation' of land. As a result, wide strips of slopes are deforested, notably at Cerro León on the eastern slope, Itatí on the western slope, some areas in the southern massif, and around Fracción and Mainumby in the north of the range. In these areas wealthy farmers of German origin from Colonia Independencia are buying land from small peasants. All the newly purchased land is immediately being stripped of its forest cover and planted with grass.

Deforestation of Yvytyrusu has serious implications for the hydrological cycle of river systems in the range, with erosion of banks, accumulation of sedimentation and eutrophication in many streams, especially in the north. Floods are more likely to occur during rainy periods, while in the dry season the same rivers disappear altogether. All activities in the valleys are potentially affected, including agriculture and cattle breeding. But the health and environmental impacts for small farmers are the most severe. Springs used by peasant families for drinking water are affected, exposing people to various illnesses, and water courses are contaminated by agro-chemicals, domestic sewage and all types of waste. For some communities, pollution has increased dramatically, affecting both people and aquatic flora and fauna.

Our surveys suggest that demographic pressures are high in the area. Of a sample of 80 families, 74 per cent owned less than five hectares, while 64 per cent have more than four children. Many landholdings are situated on steep slopes where the soil is less than one metre deep. Moreover, the high birth rate leads to continued land fragmentation. In the community of San Vicente, for example, one farmer, Felipe Areco, father of 13 children, originally had an 18-hectare lot. He has given three hectares each to his sons, Ildefonso (who has seven children) and Eulogio (six children), three hectares to his daughter Concepción (three children), three hectares to his son-in-law Nazario Machado (two children), and one hectare to his niece Brígida Silva (seven children). Similar trends are evident in other communities.

Not surprisingly, this phenomenon is stimulating migration. But the main burden of high fertility rates falls on women, as they are the ones who migrate to the cities to seek employment as domestic workers. Due to a lack of education, they generally earn low wages and have access to minimal social security. In addition, they raise their children under precarious conditions, their health affected by frequent pregnancies and lack of adequate nutrition and medical care. Women are responsible for the nutrition of the entire family. Hence, they must often improvise to find ways of obtaining food. According to the women who were interviewed, their day starts at 4:30 in the morning, when they get up, set the fire, prepare breakfast and feed domestic animals. At 7 a.m., they walk to the family plot and help the men with the agricultural chores. At 11 a.m., they return to the house, make lunch, tend the vegetable garden and work in collective economic activities such as honey production with other women. They then wash clothes, gather firewood and water, and make dinner. Their work takes 16 to 18 hours a day, without any period of rest. This is a frequent reality for peasant women all over Latin America. Women nevertheless saw a chance to form several committees to produce handicrafts.

During our work with these communities we observed no apparent discrimination by men against women, nor any explicit objection to the participation of women in productive activities, education and decision-making processes. In some instances, women were able to involve their male partners, for instance in nutrition courses and the development of vegetable gardens. Nevertheless, many women do not participate in communal activities for reasons of illiteracy and lack of experience. They may be persuaded to attend public meetings in the company of other women, but in general they maintain a passive attitude and feel inferior. Another major factor which limits women's opportunities for participation is the high number of children. Their partners are often not keen to assume responsibility for the children while the women attend meetings or courses.

To deal with these problems, the Asociación de Pobladores del Cerro Yvytyrusu is now promoting cooperative production and marketing. A community grocery store has been established in San Vicente. Beekeeping groups have been organised in Florida, San Blas, Itatí, Zorrilla and Tacuarita.

A small sugar cane processing and molasses plant is now operative in San Vicente. There are fish ponds in San Blas, San Vicente and Zorrilla. Two tree nurseries have been developed in Mainumby and in Tacuarita to produce firewood and fruit. Traditional *ao po-i* embroidery is being revived. Water provision systems have been installed in Florida and Tacuarita. All these efforts are aimed at a better management of the communities' natural resources. It must be kept in mind, however, that these communities only constitute 20 per cent of the total peasant population of the Cordillera. Most peasant families continue to live in poverty.

At the same time, there is a wealth of traditional peasant land-use practice that should be consolidated, promoted and in some cases rehabilitated. Many practices are very sustainable, including the cultivation of diverse food crops in family lots along with orchards and medicinal herbs. These traditional systems assure the family of enough food, provide remedies for many common illnesses, and often generate a small surplus for sale. In addition, appropriate technologies for the fabrication of tools and construction material such as wood, iron, adobe bricks and stonework need to be revived or developed.

External Influences

In the dominant view of development in Paraguay, the existence of forests is a sign of backwardness. Forests are seen as 'unexploited' lands that must be cleared and converted into 'productive' fields. As in Yvytyrusu, national deforestation rates have been alarming. During the late 1980s, annual forest loss reached approximately 500,000 hectares. Currently, the deforestation rate is about 250,000 hectares per year. Most of the vast forest area that less than two decades ago covered more than half of Paraguayan territory has been destroyed and replaced by soya bean and wheat fields, pastures for cattle grazing, or a patchwork of small peasant plots. None of those responsible for the implementation of this development model have taken into account the fact that these forests are essential to the environmental health of the whole region. Nor have they realised that the soils in most former forest areas are not suitable for agriculture or cattle grazing.

The fact that these forests were already inhabited by native peoples was also ignored completely. Peoples belonging to diverse cultures had developed integrated ways of life among other living species. But their land was sold to strangers with them on it, as if they were part of the wildlife. For many of them, their world – their homes, livelihoods, medicines, history, the entire universe of their symbols – disappeared with the destruction of the forest. This world was suddenly invaded by thousands of people with alien ways of living, who settled down and started to cultivate a few exotic species that soon covered the ground from horizon to horizon. Today, the surviving native forest peoples are cornered in small wooded areas that have remained after the destruction. Many died. Some migrated to distant regions, and some are exploited as labourers on the farms of the new-comers. Several indigenous communities are now struggling to obtain legal title over the small and already degraded residues of forest.

The ecological impact is clear. Hundreds of creeks and rivers draining these areas are dead, their water polluted, their beds filled with sediments transported from the eroded soils. Each year the owners of large farms need to invest more to produce less. Poor peasants cannot afford the cost of fertilisers and pesticides, and must abandon their small plots to join the growing numbers of landless. According to the National Forest Service, Paraguay will become a net importer of wood products in the next few years. The remaining production forests are no longer sustainable, since domestic demand for wood is higher than natural regrowth. Many peasant communities lack even firewood to meet their daily needs. The forests in eastern Paraguay have virtually disappeared. The Chaco,[4] which has the highest rate of deforestation, is now at risk of becoming a desert.

In the Cordillera del Yvytyrusu, the Mbya Guarani and the Ache, the most staunch protectors of their forests and their traditional ways of life, have suffered the most. A recent siege of the Ache territory in Chupa Pow by peasants is just one example of the continued persecution. In May 1996, a group of landless peasants surrounded the 7,000 hectares of forest where the Ache community lives, claiming that it was 'unexploited land' that should be eligible for agrarian reform. Yet the peasants were merely being used as pawns for Brazilian timber merchants located along the border. Even though it would be illegal, these Brazilian entrepreneurs are looking to extract all the valuable trees. The Ache people, whose lives depend on the existence of the intact forest, organised an armed defence of their territory and have vowed to defend their forest with their lives. Their numbers have already dropped dramatically from an estimated 1,000 persons in 1910 to about 300 today.

Proposals for the Protection of Yvyturusu

Based on paleoclimatic, geological, biogeographical and ecological studies undertaken by Sobrevivencia, supported by the findings of German geo-logical specialists, we can conclude that this ancient forest has unique characteristics. It is much older than the Paraná Forest which until two decades ago surrounded the Yvyturusu Forest. To protect this invaluable biological diversity, the Paraguayan government is strongly urged to enlarge the boundaries of the current national park and ensure that sufficient resources are allocated for the area's conservation. Sobrevivencia will continue to collect scientific evidence to support its plea.

The condition for these conservation efforts is that the survival and cultural integrity of indigenous communities is guaranteed. The constitution entitles the last group of Ache people who were expelled from Yvyturusu forest and transferred to Canindeyu and Caaguazy to reclaim their ancestral lands. This right must be incorporated into any management plan for the Cordillera del Yvytyrusu. Rock inscriptions, such as those found at Tororö, should be preserved as cultural monuments. Sobrevivencia has proposed that the integrity of Yvytyrusu is best secured if it is given the status of a biosphere reserve and, if possible, incorporated into a larger biosphere reserve. In this way, different parts of the range are designated under specific categories, and Yvyturusu would still receive special status.

Under no circumstances should Yvyturusu be considered as part of a managed resources area, as is now proposed by the government. We suggest that the whole southern massif, an area of some 10,000 hectares, and some 6,000 hectares in the central massif should be designated as a totally protected core zone. The Ache who were expelled from Yvyturusu should be permitted to return to the range, to the forested areas adjacent to these core zones. The remaining parts of the range – including hilltops and steep slopes, upper watersheds and water-courses, and the margins of the northern massif, peasant land and cattle ranches – can be gazetted as protected forests and buffer zones.

In general, Sobrevivencia proposes several measures to protect the remaining forests of Paraguay:

- There is an urgent need to put an immediate halt to deforestation in all the remaining native forests of Paraguay.
- Forests on all watersheds and along all water-courses must be strictly protected.
- Forests must be recognised as life-support systems, and as the traditional territory of indigenous peoples.
- The value of indigenous knowledge and forest management for society as a whole must be recognised. Legal and concrete action must be taken to ensure that sufficient and adequate land is allocated to all indigenous communities in Paraguay.
- There is a need to increase forest cover in the country with native forest species. Priority should be given to degraded watersheds, such as the San Joaquín and the ranges of Mbaracayú and Amambay. Similarly, reforestation should take place on substantial portions of all major agricultural and cattle-grazing lands. Tree species should satisfy future demand for wood and other basic needs.

The creation of a biological corridor following the river basin would offer the best chances for consolidating what remains of the Alto Paraná Forest in eastern Paraguay.

This corridor should include national parks, biosphere reserves and indigenous territories, as well as areas that require reforestation with the objective of production forestry. This corridor should include at least the following regions: the Cordillera de Amambay in Amambay and Canindeyu provinces, the Cordillera de Mbaracayu in Canindeyu Province, Sierra de San Joaquín in San Pedro and Caaguazu provinces, the Cordillera de Caaguazu in Caaguazu Province, the Cordillera del Yvytyrusu in Guaíra Province, Serranía de Monte Rosario in Guaíra and Caazapá provinces, and the Cordillera de San Rafael in Caazapá and Itapuá provinces. Preferably, it should also extend to the north to include the *cerrados* (open shrubland) and forests of Mato Grosso do Sul in Brazil, the Gran Pantanal del Alto Paraguay, and, to the south, the humid subtropical savannahs and gallery forests of Itapuá, Misiones and Neembucú, as well as Yvera in Corrientes, Argentina.

- Research on the sustainable use of natural and man-made forests that have no protected status should be promoted. This should give special attention to socially and environmentally acceptable alternatives as a substitute for massive forest exploitation to feed steel mills, the ceramics industry and the energy sector.
- An environmental monitoring organisation, based on the involvement of local communities and organisations, should be created to facilitate the protection of forests.
- Fiscal incentives should be introduced that would help generate interest in the preservation and restoration of forests.
- Rural settlements should be assisted to ensure that they respond to locally specific environmental characteristics of each site and are never created at the expense of existing forest areas, either in eastern Paraguay or in the Chaco.

NOTES

1 Of the *Saturniidae* family, some of the largest and most conspicuous species will be mentioned. Some specimens of *Copiopterix sonthonnaxi*, hitherto known exclusively from the Serra do Mar in southern Brazil, between Rio de Janeiro and Curitiba, were captured in the Cordillera del Yvytyrusu. *Dysdaemonia fosteri*, first found in Sapukay, at the foot of the nearby Cordillera de los Altos in central Paraguay in 1905, and then during the 1920s in the lower Andean altitudes between Tucuman, Argentina and Santa Cruz de la Sierra in Bolivia, also lives in the Cordillera del Yvytyrusu. *Citheronia brissotii*, widely distributed within the Serra do Mar between Rio de Janeiro and Paraná, does not exist in the rest of southern Brazil, but was also found in Yvytyrusu. Many other species of the same family of the genera *Automeris*, *Arsenura*, *Scolesa*, *Adeloneiveia* and others, show similar distributions and were found in Yvytyrusu. Two endemic species of the *Sphingidae* family will be mentioned as well. Two specimens of *Manduca fosteri*, of which specimens were found in Sapukay in 1905 and never since, were captured at the top of Cerro Acatí, Yvytyrusu in 1993. The only place *Xylophagus fosteri* is found apart from sporadic occurrence throughout the Cordillera del Yvytyrusu, is in the nearby ranges of Los Altos and Yvycui in central

Paraguay. Southern Brazil has been a well-known haunt of these two families since the middle of the last century. Studies on these families have also been carried out in Paraguay over the last 17 years. The fact that these species were not detected in any other part suggests that they are endemic and have a disjunct distribution. Of the 30 known species of the *Automeris* genus that live southward of the Amazon forest, ten are restricted to the southern region of the Andes. The remaining 20 species were all described between 1775 and 1906. In this context, the capture of several specimens of a new species of this genus in the Cordillera del Yvytyrusu in 1992 and 1993 is even more noteworthy. This species is surely endemic to this area and can only be found in little-disturbed mountaintops. This new species will be described in a forthcoming article.

2 There are two primary schools associated with the association, in San Vicente (up to the sixth grade) and in Mainumby (up to the third grade). Primary and secondary schools can be found in Melgarejo, the district capital of Colonia Independencia. There is no transportation between the communities and Melgarejo. In Villarrica, the provincial capital, some 35 kilometers from Melgarejo, there is a technical school and university.

3 Research plays a vital role in these efforts. Meetings and field trips are organised in collaboration with both the peasant communities of the Cordillera and the Mbya and Ache communities originating in Yvytyrusu. Ecological studies are conducted in the whole of the Cordillera. Furthermore, legal instruments pertaining to the status of the Cordillera del Yvytyrusu are being inventoried and analysed. A number of maps have been drafted with the help of aerial photographs and Geographical Information Systems (GIS). In addition to showing baseline data, these maps indicate the suggested zoning for Yvytyrusu Biosphere Reserve, as now proposed by Sobrevivencia. A set of audio-visual materials has been developed as an educational tool to help raise conciousness among the inhabitants of the Cordillera and the broader population. Emphasis is being placed on the importance of the Yvyturusu range for the recovery and maintenance of environmental quality in the region. The rich biodiversity of the range and the cultural diversity of its traditional populations are explained. These activities are also essential to foster a dialogue with government officials, notably with the Ministry of the Environment, Parliament, the National Commission for the Defence of Natural Resources and the Environmental Attorney's Office of the General Attorney's Office. A rapport has been built up with these bodies, in collaboration with other environmental and indigenous rights organisations, both at the national and international levels.

4 The Chaco is the second most important biodiversity-rich area after the Amazon. It covers more than a million square kilometres and extends from South Bolivia to Cordoba in Argentina, and from West Paraguay to the foothills of the Andes. The area harbours extremes in vegetation, ranging from wetlands to semi-arid forests and scrubland.

7 Living in the Atlantic Rainforest

Towards Community Intellectual Property Rights in Brazil

VITAE CIVILIS[1]

Brazil

The purpose of this study was to identify the perceptions of traditional communities and others regarding community rights to use and manage forest resources, and the use of local peoples' traditional knowledge – notably about medicinal plants – by external agents. The study focused on communities in the Juréia-Itatins Ecological Station, an area of spectacular biological diversity and residual Atlantic rainforest in the Ribeira River valley, southern Sao Paulo State. This is Brazil's most endangered biome. Originally its area covered more than 1,085,500 square kilometres, stretching from the states of Rio Grande do Norte to Rio Grande do Sul. Because of intense deforestation that began with the European discovery of Brazil in the sixteenth century, this biome currently covers just over 95,641 square kilometres – less than 9 per cent of its original size.

The ecological station was created by state law in 1987, through the efforts of ecologists, government staff and politicians. Our main purpose was to provide legal protection for this precious area and to prevent it from becoming the object of real-estate speculation. In addition to that, we were concerned about the possible construction of two nuclear power plants. Local inhabitants, meanwhile, indicated that they had never been involved in decision making about the future of their region.

The area is classified as subtropical, with alternating highly humid and semi-dry seasons. The average temperature is 22.7°C and the annual rainfall is 4.170 mm. Juréia Itatins encompasses five towns and covers some 80,000 hectares. It includes several large rivers fed by many tributaries and sources, part of the coastal mountain range, and the southern floodplains of the Sao Paulo coast. The region shows an extraordinary level of biodiversity and high numbers of endemic species – 70 per cent of the palms and bromelias, 55 per cent of the arboreous and 40 per cent of the non-arboreous species are endemic. The range of coastal vegetation varies from sand dunes, coastal plain forest (*restinga*), mangroves, coastal plain, slope forest

113

and high altitude forest. The vegetation is more exuberant and diverse on the lower mountain slopes (the slope forest) than either at sea level (the coastal plain) or at altitudes between 1,000 and 1,600 metres (the high altitude forest). In this latter kind of forest, the presence of plants associated with humid conditions – such as epiphytes (orchids, bromelias, mosses and lichens), ferns, palms and a variety of hardwoods – is remarkable.

It is important to remember that the very reason such areas remain intact is because until now they have escaped economic interest. Local people have been excluded from mainstream socio-economic and political development. These areas are preserved because communities with small populations and environmentally sensitive cultures do not seriously disrupt the ecosystem. But the exclusion of local communities from the establishment of protected areas is intensifying social tensions and legal conflicts. These are the direct result of marginalisation and social disintegration.

Permanent settlement and use of natural resources are now being increasingly prohibited inside the various categories of protected areas in Brazil. In addition, there is a general lack of support to implement systems of sustainable management which allow communities to satisfy their basic material and spiritual needs simultaneously. At this crossroads, local communities may follow the path of non-sustainable harvesting of natural resources if one denies them traditional access to resources. These same communities, equally, could become the chief guardians of the forest if they were permitted to carry out sustainable extraction of basic resources in order to survive. Consequently, efforts to identify new strategies and instruments for environmental management by traditional comununities in areas of great interest are a prerequisite for conservation.

In this study, Vitae Civilis (the Institute for Development, Environment and Peace) examined the ethical and political involvement of researchers, NGOs and other external agents. We questioned their role in providing financial or other benefits to the traditional communities, or *caboclos*, living in the so-called 'environmentally protected areas' (EPAs), in exchange for contributions by these communities to science, conservation and the use of biological diversity. Our surveys revealed that although many agree that in principle local people are entitled to financial or other benefits from the use of their knowledge, Brazil as yet has no legislation regulating the recognition and protection of intellectual property rights of traditional communities. There is, moreover, a general lack of awareness about the implications of intellectual property rights.

Environmental Legislation, Legal and Social Conflicts

Various laws regulating the use of natural resources affect the traditional livelihoods of communities living either inside or outside EPAs.[2] The

Forestry Code (Federal Law No. 4,771/65) deals with ownership, utilisation and conservation of forest resources in general, while Federal Law No. 750/93 deals with the same issues as they apply to the Atlantic rainforest. The Hunting Code (Federal Law No. 5,197/67) prohibits the hunting of wild animals, except by special permission in indigenous territories. The Constitution of Sao Paulo State (1989), however, explicitly prohibits hunting throughout the state. The Fishing Code (Law No. 221/67) introduces restrictions with regard to the type of implements and methods, seasons, species and waters used for fishing. Law No. 6902/81 deals with the creation of ecological stations and EPAs.

The Forestry Code considers the existing forests in Brazilian territory as assets of common interest for all Brazilian citizens. The creation of permanent EPAs is the strictest category of legal restriction of forest use, and is applied especially to fragile forests, hillsides with slopes over 45 degrees, and vegetation at altitudes higher than 1,800 metres. Other limitations imposed by the Forestry Code concern the harvesting of plants. This general prohibition is not as restrictive as EPA status. In fact, this clause creates a loophole that could lead to the extinction of flora species. The Forestry Code also defines 'mandatory legal reserves', minimum areas of vegetation (20–50 per cent) that must be preserved within a rural property.

Apart from these general regulations, the Forestry Code invests the state and federal governments with the power to prohibit or limit the cutting of vegetation considered to be endangered. On 10 February 1993, the President of the Republic issued Decree No. 750 on the Atlantic rainforest. Based on the Forestry Code, the decree prohibits cutting, harvesting and suppression of primary vegetation or vegetation in the middling to advanced stages of regeneration.

According to Diegues (1994), Brazil has uncritically imitated the North American 'conservationist' concept of EPAs. Created at the end of the last century, this approach is based on the belief that humans are destined to be the destroyers of nature; it therefore aims to impede and dismiss any relationship between man and nature that is not purely aesthetic or scientific in purpose. The classification and management of EPAs correspond to specific concepts of environmental protection, ranging from 'conservation of ecosystems of great scenic beauty' to those containing endangered species or biological diversity. Currently, there is increased interest in biodiversity conservation for the sake of its use in biotechnology, and exclusively in scientific research.[3]

In Brazil, these areas fall into two categories. The 'direct use' areas allow for human habitation, as is the case with 'extractive reserves'. The 'indirect use' category includes parks, biological reservations and ecological stations. All are subject to permanent conservation, and are meant to serve scientific, cultural, educational and recreational purposes. They are established and managed by federal, state or local governments. Ecological stations are areas representative of Brazilian ecosystems. They are set aside for the purpose of protection, basic and applied ecological research, and the development of conservation education. Ninety per cent or more of the area is supposed to be integral to conservation of the biota. The remaining area can be used for ecological research, which may modify the natural environment, provided that it is authorised and does not risk the survival of existing species. Ecological stations are by definition a public domain, and their creation thus often involves the expropriation of private property.

The law prohibits herding domestic animals, harvesting natural resources and carrying arms or nets inside ecological stations, although it does not explicitly forbid the presence of human settlements. Still, these restrictions seriously hamper the livelihoods and culture of local communities. Many parks and ecological stations, such as Juréia-Itatins, were established in areas inhabited by rural and traditional peoples, creating severe and diverse problems not only for communities but also for the responsible authorities. The bill on a National System of Environmentally Protected Areas, presented to the National Congress in 1992, is yet to be approved after much study and debate during the years that followed – although it has broadened the discussion of environmental conservation in Brazil (see Table 7.1).

Table 7.1 Current legal status of communities under different Brazilian EPA regimes, compared with their status under EPA proposals in the National System of Protected Areas (SNUC) Bill, November 1995

	Existing EPAs	EPAs proposed by SNUC
EPAs	X	X
Ecological stations		
Natural monuments		Depends on responsible managing authority
National park		X (5% of the land)
Wild life refuges		X (5% of the land)
Biological reserve		
Extractive reserve	X	X

Source: National System of Protected Areas (SNUC) Bill, November 1995.

The Juréia-Itatins Ecological Station is managed by the Forestry Institute, a part of the Sao Paulo State Department for the Environment. This agency is responsible for the establishment and management of all the state's EPAs, with no participation by local inhabitants. Yet the new regulations drastically curtail local peoples' user rights over the forest. With no alternative source of income and facing precarious living conditions, an increasing number of families have migrated to the cities. Those that remain are often forced to engage in illegal activities such as harvesting palm heart (*Euterpe edulis*) and *caxeta* wood (*Tabebuia cassinoides*). Inevitably, conflicts among local inhabitants, the government and ecologists have intensified. Local inhabitants have also started to organise politically through the Union dos Moradores da Juréia (Union of Inhabitants of Juréia) and the Union dos Jovens da Juréia (Union of the Youth of Juréia), with the aim of seeking solutions for their problems. Some citizens' groups support these unions.

The People of Juréia

The ecological station is inhabited by 365 families – 1,531 individuals – in 22 communities, who have lived there long before the establishment of this protected area.[4] The main communities are: Cachoeira do Guilherme, Aguapeu, Paraia do Una, Grajauna, Rio Verde, Rio das Pedras, Serra and Rio Una. The people are descendants of white, black and indigenous frontiersmen, or *caboclo*, and are usually described as hillbillies (*caipira*), other hill dwellers (*capuava*), beach dwellers (*caieara*) and riverside dwellers (*ribeirinho*).

According to the law governing EPAs, the lands are owned by the government. A 1991 government land ownership survey determined that almost all families in Juréia-Itatins are squatters, caretakers, servants or sharecroppers. Only a handful of families owned land. In reality, however, the majority of land in the area is privately occupied. The families' living conditions are often dismal. For instance, there are no sewage treatment facilities, septic tanks or other sanitary facilities. There is only one health centre in the entire area. The few schools are often closed owing to lack of facilities and teachers. Houses are constructed with wood and other materials obtained from the forest and there is no electricity. Water is taken from waterfalls, rivers and wells.

The main sources of food and cash income are small clearings in the forest planted with rice, corn, beans and manioc. In addition, families engage in the collection of plants, hunting and fishing. Under the current legislation, the area under cultivation has been reduced drastically. It has become difficult to obtain the licences required to extract wood, particularly for making canoes. At the same time, illegal and unsustainable methods of harvesting forest produce, such as palm heart and game, have intensified. Those caught face heavy penalties of up to three years' imprisonment. More income must now be derived from pensions, off-farm jobs and employment with conservation area authorities.

A great variety of products is obtained from the forest. Wood is used for the construction of houses and the manufacturing of utensils, and for canoes, boats and wagons. Vines, grasses and clay are collected for thatching and various other uses, such as food, medicines, musical instruments and domestic utensils. Prior to the establishment of the ecological station, surplus products were sold. Hunting provides families with a protein supplement to their daily diet, and plays an important role in social organisation. Riverside dwellers and beach dwellers engage in small-scale fishing. Shifting cultivation, or *coivara*, is the most dominant agricultural practice. Forest is cleared by cutting the larger trees and burning. The clearing is used for four to five years, then abandoned to allow the soil to recover.

Research by Vitae Civilis revealed that, in spite of restrictions imposed by Brazilian environmental legislation, the relationship between the communities and the natural environment remains fairly close and harmonious. Local people still possess remarkable skills, such as making meteorological predictions for the short term on the basis of temperature, the wind's direction and intensity, and the observation of animal behaviour. Following their own logic, local people can identify, classify and judge potential uses of flora and fauna with great accuracy. They adhere to certain rules which guide, for example, hunting procedures.

The traditional communities have achieved a high level of social organisation and cohesion under the spiritual leadership of Satiro Tavares da Silva (1927–1995). Satiro was 16 years old when his family arrived in the Cachoeira do Guilherme community from another region in the Ribeira Valley. He assumed leadership of the community following the death of his father, the former spiritual leader. He and his wife Alice lived next to the spiritual centre where Satiro conducted wedding and baptism ceremonies, and attended to the sick. Congregations in which all communities participated were held on the first Saturday of each month. Before his death, Satiro did not transfer authority to anyone – although his son now performs an important role in the comununity – explaining that all members of the traditional communities in Juréia-Itatins now share responsibility for the continuation of the spiritual work. In the face of the hardships that have followed the establishment of the ecological station, Satiro's leadership has been vital, and his influence continues. Religion is the strongest cultural element affecting the communities' relationship with nature. Some diseases, for example, are often interpreted as demonic possession, and a religious ritual and traditional herbs are used to exorcise the troublesome spirit. Only those endowed with these skills preside over such rituals, a fact which tends to legitimate the spiritual leader's hegemonic role in the community.

The accumulated ethnobotanical knowledge existing within the communities is enormous. People have developed a great variety of techniques to obtain trees and plants and to make use of their ingredients and characteristics. This knowledge is passed on from generation to generation. Plants are harvested in nature under specific conditions and processed in different ways. We found a total of 1,048 different applications (part of the plant, purpose, time and method of harvesting) for more than 400 different species. As regards ethnopharmacological knowledge – the medicinal use of natural resources – those interviewed mentioned 147 different illnesses that are treated with plants or other natural ingredients. These are prepared in different ways and administered in specified doses for a determined period of time and with certain precautions. We recorded the use of 308 plant species and 1,131 therapeutic methods for 141 illnesses. Of these 308 species we were able to identify some 200, of which 124 are endemic and 76 exotic. Our findings concurred with the results of 158 pharmacological studies mentioning about 80 of these endemic species, and 36 mentioning about 50 of the exotics. The studies were conducted over the last 30 years.

Although the communities continue to use their traditional knowledge about natural resource management there is a risk that this knowledge is disappearing as attitudes towards the forest change. Partially as a result of the new restrictions imposed by the protected area, young people are forced to leave the region in search of work in neighbouring towns. Consequently we are facing a number of complex dilemmas. How can the communities improve their standard of living? How can local communities participate in the management of their natural environment? How can their knowledge benefit humanity while at the same time guaranteeing that communities enjoy intellectual property rights and receive (financial) compensation?

Some important initiatives and precedents motivated us to address this challenge. First, a number of extractive reserves have been created. Second, the Convention on Biological Diversity formally recognises the right of indigenous and traditional peoples to receive benefits in exchange for their knowledge of the management and utilisation of natural resources. Third, legislative proposals on the National System of EPAs and the protection of industrial patents have been introduced in the National Congress.

User Rights and Intellectual Property

Since 1989 Vitae Civilis has been working in the Atlantic rainforest area of Sao Paulo State. Our objective has been fourfold: to contribute to environmental conservation and sustainable local development; to diagnose emerging problems and seek solutions in partnership with local people; to develop scientific standards for the use and management of natural resources; and to contribute to the empowerment of local populations and the improvement of their standard of living. These local activities have allowed us to gain experience and data that can be used in the dialogue on global policy issues. The case of Juréia shows the direct correlation between local and global issues.

The study used data from action research conducted between 1989 and 1995. Detailed information was available from ethnopharmacological and ethnobotanical surveys and from research into the socio-economic and cultural conditions of the traditional communities. The aim was to obtain a profound understanding of how the natural environment is used, and to assist comununities to improve their sources of income and to obtain better medical care.[5]

To deepen our understanding of the various perceptions of comununity intellectual property rights, Vitae Civilis approached a whole range of individuals and groups concerned about environmental protection. In our discussions it became evident that the representatives of different government institutions – such as the Sao Paulo State Department of the Environment and its Forestry Institute – held the view that nature must be protected against man. The communities, however, opposed this view. They did not believe that conservation measures are necessary, as they do not recognise the threats to the forest. They argued that 'when our ancestors arrived here 200 or 300 years ago, everything was as it is now. It is all nonsense, this ecology story.' Almost all the inhabitants mentioned losses or damage to livelihood as a result of the creation of the ecological station, and they denied having caused any ecological degradation.

The socially oriented NGOs stressed that conservation measures should be dealt with in relation to social problems. They referred to the approach of the State Department of the Environment as 'authoritarian', and felt that the restrictions on the use of natural resources are overly rigid. They agreed that the establishment of the Juréia-Itatins Ecological Station has threatened the population's source of basic necessities.

Academic and research institutions felt that biodiversity conservation should be accompanied by scientific research. But conservation organisations agreed with state bodies that biodiversity conservation should be conducted in a systematic manner, and consequently, should be the responsibility of the state. They both argued that the local population was at risk of being overwhelmed by real-estate speculation, whereas the creation of the ecological station halted this process. The fact that the new protected status imposed restrictions on local people was viewed as an unavoidable consequence.

All those we spoke to affirmed local people's rights to remain in the area and to use the natural resources. While the government representatives and conservationist NGOs insisted that only traditional communities should remain – albeit with certain limitations – the socially oriented NGOs felt that communities should be allowed to stay regardless of whether they were defined as 'traditional'. They considered people's use of natural resources as compatible with conservation. The academic interviewees declared that the slogan 'land for those who need it' should not be applied to the protected areas, but rather to large estates. The communities, meanwhile, spoke of their bond with the region as a 'love for the land', a sentiment that keeps them there, in spite of the restrictions they have to suffer.

Government agencies and conservationist NGOs shared the opinion that the use of natural resources should be linked to rules for environmental protection. Since there is no technical information on this aspect available, the state should promote research with the participation of communities to establish guidelines. The academics supported this stand, adding that efforts are necessary to improve the legislation of protected areas. But socially oriented NGOs emphasised that communities were already in the area before the protected area status was declared. Moreover, they did not consider the methods of resource use to be detrimental to the environment. Hence they felt that any definition of sustainable use should be discussed with the communities, thereby taking into account their extensive knowledge of the environment.

In summary, most people interviewed believe that local people should be allowed to stay in the area, but felt that their standard of living needs to be improved, that land ownership must be better regulated, and that infrastructure and basic facilities should be upgraded. To this end, the protected areas could be transformed into a 'mosaic' combining various land use types and functions – such as hunting for subsistence, rearing wild animals, small-scale agriculture, and various other sources of income for the communities. Several of these suggestions, however, would require legislative or constitutional amendments.

Regarding the communities' intellectual property rights, the inhabitants of Juréia-Itatins tended to be unaware of the importance and potential of their knowledge about the environment. They believe they have little of value to offer modern society, and it is far from clear to them how they could possibly benefit from their knowledge. The communities were more concerned with their practical welfare than with the abstract concept of 'knowledge as property'. In contrast, the other groups interviewed by Vitae Civilis had no doubts about the great importance of community knowledge, especially regarding traditional medicine, which was considered to be of outstanding potential value to the larger society. But communities would need to be properly compensated for their traditional knowledge. Various people pointed to inadequacies in current legislation for protecting traditional knowledge. They mentioned the frequent difficulty of identifying the precise origin of such knowledge, as traditional learning is often collectively developed by several persons or communities.

Considerations and Proposals

Traditional knowledge is no less scientific, nor are its results less real than those of modern science based on observation and experimentation (Diegues 1994). According to the World Health Organisation (WHO), about 80 per cent of the world's population uses plant derivatives as remedies for illnesses or as ingredients in other products. Moreover, 75 per cent of the plants that supply active ingredients for drugs are estimated to have been discovered from traditional medicines used by healers in indigenous communities (Farnsworth *et al.* 1985). It is thus evident that by contributing their collective transgenerational knowledge about conservation and use of natural resources, these communities are a key to the survival of the human species. It goes without saying, however, that their knowledge and cultural identity also need to be respected. Although some may integrate into mainstream society, this does not imply that they must act and live according to the country's dominant culture.

It is obvious that traditional knowledge assumes an economic value in capitalist societies. Because of the potential of knowledge to generate wealth, economic exploitation of this asset by entrepreneurs without some form of benefit to its owner is a form of illicit or unjust gain. Judicial systems in most countries prohibit this. That the value of traditional peoples and their knowledge of biodiversity conservation is not acknowledged can never be an excuse. One cannot maintain a utilitarian vision of such communities as exploitable natural resources. Hence, sustainable development depends not only on the conservation of biodiversity, but also on respect for cultural and ethnic diversity. There is an urgent need for legal means to address the dilemmas posed by the use of community knowledge by entrepreneurs and other external agents. Otherwise, communities will simply disappear under the burden of poverty and marginalisation.

In contrast with the current bill, the newly proposed Federal Bill of the New Indigenous Statute, PL 2057/91, includes an entire section (Chapter 11) on the subject of the intellectual property rights of indigenous communities. Access to and use of traditional indigenous knowledge requires the community's prior written consent. There is still, however, no Brazilian law that covers Section (j), Article 8 of the Convention on Biological Diversity, which recognises the value of the intellectual assets of non-indigenous traditional communities. Bill 306/95, which is being introduced in the Federal Senate, deals with the control of access to Brazil's genetic resources. This proposed law contains a section dedicated to the protection of traditional communities' knowledge. It affirms that local communities have the right to receive benefits collectively for their traditions and knowledge, and to be compensated for the conservation of biological and genetic resources. If approved, this bill will have far-reaching consequences because, unlike the New Indigenous Statute Bill, it will also apply to non-indigenous forest inhabitants, 'frontiersmen' such as riverside and beach dwellers, hill dwellers and hillbillies.

In April 1996 the National Congress approved a patent bill that protects industrial intellectual property rights. Its provisions reflect the extent to which the Brazilian government bowed to pressure exerted by the US government and transnational pharmaceutical and chemical companies. This law will even permit patenting of modified micro-organisms, but it does not refer to the rights over intellectual resources of indigenous or traditional peoples. The only new element as far as the intellectual property rights of traditional communities are concerned is that pharmaceutical products and manufacturing processes can now be patented, since they are no longer included in the list of non-patentable articles (Hathaway 1994).

Although no law forbids it, it seems to us that it remains difficult to obtain patents on the inventions of non-indigenous traditional comununities. We believe that it may prove impossible to satisfy all the requirements for patents demanded by the current system. The premise of the legislation is that inventions are patentable, but discoveries are not. Inventions are considered creations – the fruit of human intelligence using natural forces for the solution of a problem that aims to satisfy practical or technical necessities. A discovery, on the other hand, results from the application of intellectual capacities to the study of natural phenomena and laws. It is not the action of the inventive spirit. It is the result of speculation and the ability to observe. This suggests that both traditional knowledge and scientific discoveries should fall outside the protection of the current patent system. According to Brazilian law, a company producing medicine based on research with traditional communities can request privileges on its invention, while the community on whose knowledge the invention was based cannot do so. Thus the community is unlikely to receive benefits, while the company will profit financially.

Another difficult problem is the definition of ownership and the question of who created the invention, as we are usually dealing with diffuse learning. Non-indigenous traditional communities are experiencing radical changes in their culture – with the result that their knowledge is spread and exported to other communities. This makes it complicated to determine the keeper of the knowledge or the 'father' of the invention. Hathaway noted in 1994 that the intellectual property rights of indigenous and traditional communities deserve a specific legal model, one different from the Brazilian patent laws because applicable to collective innovations gathered over long periods, sometimes over generations. He also argued for a law regulating access to genetic resources that would require contracts with guarantees of returns to communities through payment for information and material as well as through ensuring community participation in benefits derived from the sale of products.

Thus the value of and right to the intellectual property developed by these communities over time must be recognised. The basis of entitlement to this right is that the knowledge is the fruit of the community's intelligence, powers of observation and efforts in developing and reproducing this knowledge through generations. Recognising the right means having respect for traditional knowledge as a 'legal asset' with definite ownership. The next step is legislation protecting this asset. Laws are political instruments that regulate human relationships with the objective of eliminating inequalities; placing the strong and the weak on the same plane. Such legislation should interfere, therefore, in the relationship between traditional communities and industry, in order to ensure that benefits are fairly divided. These are recommendations of the Convention on Biological Diversity that Brazil signed and ratified, but has still to put into practice.

A possible mechanism for reducing the exploitation of traditional communities is bilateral agreements between communities and companies that regulate the relationship between the parties. Such agreements should ensure some form of redress to the suppliers of knowledge used for commercial purposes, and should guarantee the rights of traditional communities to share the benefits of co-ownership of patents and inventions. Other benefits can be anticipated in such contracts, bearing in mind the contractual principle of equality of both parties, as long as such arrangements do not jeopardise the public interest. At present there is no legal redress for breach of contract. The issue of contract duration also needs to be included. Because these communities are obviously unprepared to deal with such arrangements, legislation should impose basic mandatory clauses stipulating minimal guarantees in the contractual relationship to prevent any type of exploitation of these communities. Such legislation should also be the subject of broad discussion among the communities, academics and other interested parties. Only with such legal protection will the culture and knowledge of traditional communities stand a chance of survival. Just reward for their knowledge and other contributions to the larger society may also help to improve community self-esteem and strengthen their dignity. To our knowledge Brazil still lacks specific legislation recognising the intellectual property of communities as a legal asset. This means that there is as yet no legal framework guaranteeing such communities the right to dispose of their intellectual property as they see fit, or ensuing fair compensation for the commercial exploitation of such property.

Finally it is worth remembering that this discussion originates from the premise that these assets have an economic value. Yet for the internal dynamics of the indigenous and traditional communities, the value of such knowledge is foremost in its practical and immediate use in daily life. The majority of these populations do not live under the aegis of the capitalist system, although they are surrounded by it. The objective of the proposed mechanisms is to regulate the relationships between these communities and the capitalist world.

Brazil now faces the challenge of supporting the role of traditional communities in biodiversity conservation and natural resource use. We therefore propose that the following policies and concrete measures be introduced:

- Categories of EPAs need to be defined that give equal importance to environmental conservation and sustainable development of traditional peoples.
- Prior to the creation of such areas, studies need to be conducted in partnership with communities which live within or adjacent to them.
- It is crucial to establish partnerships in management in order to share the responsibility for environmental protection with the various interested parties. This requires managing councils for each protected area, based on equal representation of government, NGOs, scientists and communities.
- It is time for innovative and exemplary projects of biodiversity conservation which involve the practices of traditional communities.
- Government efforts to conduct studies and activities aimed at strengthening practices of traditional communities within protected areas should be supported. Such actions range from reclassification of the areas to changes in official management policies and practices, and the evaluation of innovative models of participatory management.

The path by which traditional communities develop their knowledge and inventions is different from that followed by research projects in industrial laboratories that generate patent rights. More studies are required to address the various unresolved legal, anthropological, ethnological and political aspects, in order to identify formal options that will both safeguard traditional knowledge and regulate the use of such knowledge by society. It is, however, worth emphasising that economic or social returns to the proprietors of this knowledge presuppose a prior understanding that compensation is right and just. Hence, it is essential that external agents who relate to the traditional communities should make a political and ethical commitment that leads in that direction.

NOTES

1   Vitae Civilis's technical team consisted of the following persons: Gemima C. Cabral Born, Rubens Harry Born, Oriana Aparecida Favero, Cássio Noronha Inglez de Sousa, André Rodolfo Lima, Sandra Pavan, Patricia Vieira Sarmento, Patricia Alvarenga, Paulo Sérgio Stockler, Cynthia Regina Caly, Ivanney Pessoa Moreira Martins (adviser), Cristina da S. Machado, Andréia D. Fornazier, Dauro Marco Prado, Sidioney Onzio Silveim (administrative support). Vitea Civilis wishes to thank the following people: Satiro Tavares da Silva, spiritual and political leader of Juréia, the traditional communities of Juréia-Itatins Ecological Station, and all the interviewees and their respective organisations for supplying information and perceptions; Patricia C. F. Alvarenga and Paulo Sérgio Stockler, Ivanney P. M. Martins, Sidioney Onezio Silveira, Cynthia Regina Caly and Daniela Milstein for voluntary help; António Carlos Alves de Oliveira for valuable information; Humberto Mafra of the Fundacion Francisco and Paul Wolvekamp of Both ENDS for great help in obtaining financial aid; Both ENDS and the Netherlands Committee of the International

Union for the Conservation of Nature (IUCN) for financial support for this study; the Rainforest Alliance (New York) for financial support which enabled us to develop the Juréia project upon which this research is based; the Rainforest Action Network, especially José Roberto Borges, who also contributed to developing part of the Juréia project; Simone Bilderbeek Lovera, Miguel Lovera and Wouter Veening of the Netherlands Committee of the IUCN; and, last but not least, our companions, friends and families for their moral support and understanding during our absence.

2 Brazilian laws governing the use of natural resources apply to all citizens. The section on the environment in Brazil's 1988 Constitution raised the concept of 'ecologically balanced environment' to the status of 'an asset of common use for the people and essential to a healthy quality of life'. Such an asset is then defined as 'public property destined for the use of all citizens without distinction, without anyone's exclusive or privileged use of the asset', or as Hely Lopes Meirelles (1993) notes, 'each individual has equal rights to benefit from the asset or to deal with burdens resulting from it'. General regulations of public order are applied to the use of these assets, ensuring hygiene, health, morality and good habits, without specifying persons or social categories. The asset of 'common use' is considered indisposable and, therefore, non-negotiable.

3 'Subsidies for Discussion', paper at Workshop on Political Guidelines for EPAs, World Wildlife Fund (WWF), November 1994.

4 Sao Paulo State-SMA/CPRN/IF/ELS, 'General Census of the Inhabitants of the EEJI' (November/December 1990), 1991.

5 We tried to help the communities by contributing research data on the cultivation, processing and sale of medicinal plants, and a community health centre was established, staffed by a health agent. A number of communal activities were undertaken to strengthen the community spirit and collaboration was started with the government's Health Department. As a result of these actions a number of concrete successes were accomplished. A member of one community was selected to be trained and later on was hired as a Community Health Agent. Two communities have started to cultivate medicinal plants which they sell. And, perhaps most importantly, the communities have become more involved and outspoken about developments in the region and their own priorities.

BIBLIOGRAPHY

Born, G. C. C., Diniz, P. S. N. B. and L. Rossi (1989) 'Levantamento Etnofarmacológico e Etnobotânico nas Comunidades da Cacheira do Guilherme e parte do Rio Comprido (sitio Ribeirao Branco – sitio Morrote de Fora) da Estaçao Eçológica de Juréia-Itatins, Iguape'. Consultant's report presented to the Department of Environment, State of Sao Paulo.

Born, G. C. C., Favero, O. A. and L. Rossi (1992) 'Ethnobotany and Conservation of Cultural and Biological Diversity in the Atlantic Rainforest Region (Sao Paulo, Brazil)', paper presented at the Third International Congress on Ethnobiology, 10–14 November, Mexico City, Vitae Civilis–Institute for Development, Environment and Peace, p. 8.

Born, G. C. C., Favero, O. A. and S. Pavan (1994) 'Ethnobotany and Community Development in the Atlantic Rainforest (Brazil)', paper presented at the Fourth International Congress of Ethnobiology, November, Lucknow, India, Vitae Civilis–Institute for Development, Environment and Peace.

Born, G. C. C. (1995) *Sátiro às Avessas. Um Lider Espiritual e Politico da Juréia*. Department of Anthropology, Faculty of Philosophy, Literature and Social Science, University of Sao Paulo, 36 pp.

Born, R. H. (1991) 'Aspectos conceituais, ambientais e de saúde publica do aproveitamento (reuso) de águas residuárias no solo como administraçao da qualidade ambiental', Masters dissertation, Faculty of Public Health, University of Sao Paulo, 278 pp.

Camara, A. I. G. (1991) *Plano de Açao para a Mata Atlântica*, LTDA Publishers, Sao Paulo, 152 pp.

Diegues, A. C. (1993) *Populaçoes Tradicionais de Unidades de Conservaçao: O Mito da Natureza Intocada*, NUPAUB/USP Series, documents and research report No. 1.

Diegues, A. C. (1994) *O Mito Moderno da Natureza Intocada*, Core Support for Research on Human Populations in Humid Areas of Brazil, University of Sao Paulo.

Diegues, A. C. and P. J. Nogara (1994) *O Nosso Lugar Virou Parque*, Core Support for Research on Human Populations in Humid Areas of Brazil, University of Sao Paulo.

Farnsworth, N. R., Akerele, O., Bingel, A. S. and D. D. Soejarto (1985) 'Medicinal Plants in Therapy', *Bulletin of the World Health Organisation*, Vol. 63, pp. 965–98.

Favero, O. A. and Born, G. C. C. (1994) *Desafios e Controvérsias de Métodos em Etnociências*, Vitae Civilis–Institute for Development, Environment and Peace, p. 23.

Geertz, C. (1978) *A Interpretaçao das Culturas*, R. J. Zahar ed., Rio de Janeiro.

Giannini, I. V. (1991) 'A Ave Resgatada: "A Impossibilidade da Leveza do Ser"', Masters dissertation, Faculty of Philosophy, Literature and Social Science, University of Sao Paulo.

Hathaway, D. (1994) 'Diversidade e Garimpagem Genetica', paper presented at the International Meeting on Eco-Social Diversity: Cooperation Strategies between NGOs in the Amazon, Belém, 13–16 June.

Meirelles, H. L. (1993) *Direito Administrativo Brasileiro*, eighteenth edition.

Posey, D. A. (1986) 'Introduçao-Etnobiológica: Theoria e Pratica', in *Suma Etnobiológica Brasileira* (Org. B. Ribeiro), ed. Vozes, Sao Paulo, pp. 15–25.

8 'Washing Hands with Soil'

Restoring Forest and the Land in Costa Rica

MIGUEL SOTO CRUZ

Arbofilia, Costa Rica

Costa Rica, wedged between the Caribbean Sea to the northeast and the Pacific Ocean to the southwest, is bordered to the north by Nicaragua and to the southeast by Panama. A series of volcanic mountain chains runs from the north to the southeast, splitting the country in two. On either side of the volcanic central highlands lie coastal lowlands. The smooth Caribbean coastline is 212 kilometres long and has year-round rain, mangroves, swamps, an intercoastal waterway, sandy beaches and small tides. The much more rugged and rocky Pacific coast is 1016 kilometres long, with various gulfs and peninsulas (Rachowiecki 1991).

Costa Rica hosts an enormous biodiversity. More than 200 different mammals, half of which are bats, have been recorded (among which are the sloth, the jaguar and the tapir) and over 830 bird species (including 15 parrots, the toucan and the harpy eagle). At least 35,000 species of insects are found in the country (Janzen 1983), not to mention a great variety of reptiles and amphibians, including crocodiles and both marine and fresh-water turtles (Savage and Villa 1986). The floral biodiversity includes some 9,000 species of vascular plants, 1,500 different species of orchids and 900 tree species. There is a great range of different fruit trees with, for example, several dozen species of fig (Hartshorn 1983).

Although it is a small country, Costa Rica offers an enormous variety of habitats, each with particular associations of plants and animals. Twelve tropical 'life zones' can be distinguished, with dry, moist, wet and rain forest in tropical, premontane, lower montane, montane and subalpine areas. This categorisation is based on the Holdridge system of ecological 'life zones', with each life zone hosting several different ecosystems (Holdridge, 1967). Large parts of Costa Rica are situated in the moist or wet-forest life zones. On the western slope, from northwestern Costa Rica northward, most of the natural vegetation is categorised as dry-forest zones. Herrera and Gomez (1993) distinguish no fewer than 56 different biotopes in Costa Rica.

Costa Rica is a country well known for its efforts in establishing national parks, which now comprise about 11 per cent of the total land area. But it is also experiencing increasing deforestation. Though various buffer zones and forest reserves are claimed as protected land, increasing the total area to about 27 per cent, these areas still allow logging and other exploitation. Technically a democracy, the country is also known for its centralist politics, a complete lack of mechanisms for direct democracy and – predictably, perhaps – systemic corruption.

This political profile has had an enormously adverse impact on the country's forest estate. Costa Rica has suffered some of the highest deforestation rates in the world. Some hardwood species have been logged to the brink of extinction. Other rare timbers have been exploited to the point of severe genetic erosion. During the 1970s, tracts of original forest in the Central Pacific became totally isolated islands amidst burned-out terrain. Some 15 years ago Arbofilia began to work towards 'maximum use of biodiversity for environmental and economic restoration'. Although the general situation has not changed much since, some important areas are recovering. With the regeneration of natural forest, water sources recover too. The principal mechanism of recovery for these areas is the restoration of natural biological mechanisms similar to those in the original forest.

Scene Setting

The farmers' cooperative Arbofilia is based in the mountainous Central Pacific region. Some 5,000 families live in the area, and a total population of 400,000 people along the Pacific coastline. The entire area was once covered with natural forest. Up to 1950, 85 per cent of the primary forest remained. Today, the landscape is characterised by remnants of forest, separated by an eroded environment that has become dominated by teak (*Tectona grandis*) or gmelina (*Gmelina arborea*) monocultural plantations.

The tropical fringe is situated at latitude 10°N, longitude 87°W. It is a zone of many microclimatological contrasts, with elevations from the coastal mangrove forests of the Pacific to mountains with an altitude of more than 2,000 metres. The annual rainfall is between 1,800 and 5,000 mm. The mountain area of the Central Pacific of Costa Rica harbours such a spectacular biodiversity – among the greatest in the world – because it contains multiple transitions. The area forms a linkage between the tropical rainforests of the Choco of South America and the deciduous forest of Mesoamerica to the north. From east to west the transition goes from the montane forest to mangroves on the Pacific Coast. In addition to this confluence of several types of transitions the area is characterised by a very complex mountainous topography. This has created an enormous variety of microclimates promoting a high diversity of life forms.

The study area is near the Carara Biological Reserve, a protected area of 4,700 hectares, situated in the transition zone between tropical rainforest in the south and deciduous forests to the north. Three main natural forest types can be distinguished. First there is tropical rainforest in the Biological Reserve of Cacara, with transitions to dry forest in the north. Second, there is montane and premontane forest in the La Potenciana Mountains. Third, there are the small forest areas in La Cangreja Reserve. There are also hundreds of small, isolated forest patches distributed throughout the region.

Forestry Policies

Costa Rican governments have pursued the rapid transformation of some of the most diverse forest ecosystems in the world into homogeneous plantations of one or two exotic species, basically teak and gmelina. These forestry policies were defined without broader consultation, although only in the field is the real impact known and felt. In fact, there is an enormous dichotomy between the message of the political elite and what has really happened in the country. Thus fashionable images of conservation and development have been presented assiduously at international fora, aiming to attract as much international environmental funding as possible. Loans, investments and 'green funds' have been used under the pretext of re-afforestation and CO_2 sequestration to invade the countryside with monocultural plantations. Sadly enough, the implications of these policies have not been well understood at the political level. The replacement of natural forest by monocultures is confused with 'sustainable development'. The expansion of monocultural forestry is only one of the many markers on the road towards increasing uniformity of landscape and culture.

The forestry and agricultural production systems that currently dominate are responsible for this pervasive loss of biodiversity. Deforestation and forestry practices have also resulted in serious shortages, to the extent that some regions must now import wood from elsewhere to satisfy the most basic needs. The main objective of this forestry is to use the land and the local labour force to produce raw material for export to the rich industrialised countries. Since the subsidised monocultures cannot satisfy international and domestic demand, there is great commercial pressure on local forests. This represents a serious threat to small remnants of natural forest.

Loss of Biodiversity

The reduction of the ecosystem's biodiversity is translated into a reduced capacity to renew and recover, making it more vulnerable to external changes. The enormous scale of the weakening of ecosystem vitality, in turn affecting both agriculture and forests, limits possible economic development and improvement of the quality of life for the majority of inhabitants, both in the Central Pacific region and in the country as a whole. If we compare the situation today with earlier systems of land use in Costa Rica, we can appreciate that the impact on the environment of indigenous cultures was imperceptible. This was partly because of low population density, but mainly because of a pattern of use that demonstrated respect for the variety of ecosystems. This made the impact light, distributed between several ecosystems with just a small tendency towards modification of the composition and structure of the forest. In this pre-Columbian period the impact never reached 1 per cent of the whole landscape. Later, with the arrival of the Spanish colonisation, the establishment of *chacras* (small rural and family holdings) was based on open-field agriculture instead of forest. During more than 400 years, however, no more than 15 per cent of the original forest was disturbed

It is important to note that the massive destruction of forests has taken place in the last 30 years. Deforestation has been promoted by successive governments to extend cattle raising for beef consumption, with incentives granted to burn the natural forest. The system of granting land titles required that claims to land ownership were substantiated with proof that the land was occupied and worked. This triggered unprecedented clearing of forest. Thus political and economic forces have promoted the erosion of ancient indigenous cultures, as well as the systems of traditional agriculture and agroforestry of the past four centuries. This situation has resulted in the widespread elimination of communal or tribal systems of forestry protection, including valuable knowledge about the use of local biodiversity.

In the Central Pacific the most evident consequences have been soil erosion, the corresponding impoverishment of families living in this zone, and a massive disruption of water systems, affecting people, animals and agriculture. No other ecosystem has suffered greater damage than the fresh water sources, which supply crucial nourishment for local protein production. Over recent years farmers have observed climatic changes, characterised by longer dry seasons with less rainfall and torrential rains concentrated in fewer days during the wet season. Such extreme weather patterns have provoked both floods and water scarcity.

The prevailing political system offers little direct democratic space to debate current forest policies and to bring in the options and priorities of local communities. The local perspective is only advocated by politicians to the extent that it serves the purpose of attracting foreign aid, while in reality the encroachment on the landscape of monocultures has led to the displacement of the rural population. The majority of these migrant farmers, peasants and forest dwellers will, sooner or later, end up in city slums.

Commencing Restoration Work

Considering that the government's land use and forest policies were unlikely to change in the nearby future, Arbofila decided to use what space was available to develop an ecological and cultural restoration process in collaboration with farming communities, starting in the Carara region. In 1987, Arbofilia created a map with a strategy incorporating biological principles to reunite important ecosystems. At the time such an approach was unheard of and it was difficult to find external support for the recovery strategy. The conservation model in force, taken over from the United States and Canada, had become the most important obstacle to realising the aspiration of the local cultures to take responsibility for management of the environment. Instead, the radical conservative hands-off approach left 'green isles' of conservation untouched, surrounded by poverty and increasing degradation. Although state policies continue to inhibit and restrain a people-based approach, the Arbofilia restoration process has now won enough ground to survive until environmental policies are based on equality and respect for the space of other life forms, or until local communities find mechanisms to protect themselves against inherent centralist corruption.

Recovering the Water Sources

Environmental conditions limit the real development of communities, and nothing makes this so evident as the destruction of drinking water sources. One of the priorities of the fieldwork of Arbofilia, therefore, has been the study and recuperation of water sources. This in turn has required the regeneration of forest vegetation on crucial watersheds and erosion control on adjacent agricultural land. The success of these restoration efforts boosted the enthusiasm of the communities to take up restoration work in other parts of the river basin. It is beyond the capacity of a rural organisation like Arbofilia to work at the level of the river basin, however, since this would entail land reforms, zoning policies and major investments. The required political environment is absent. Given these limiting circumstances, Arbofilia chose a strategy composed of the following essential elements:

- Maximum use of local biodiversity for ecological restoration.
- Design of productive systems analogous to the natural forest that bring direct economic benefits to the communities.
- Strengthening of the organisational skills of the communities, beginning with participatory diagnosis and leading to design and implementation of plans for ecological and cultural recuperation.
- Increased self-reliance and reduced dependence on external resources.
- Establishment of autonomous cultural space to reduce inter-ethnic conflicts and vulnerability to political manipulation.

- Cultural restoration with equitable participation of women, as a social prerequisite for environmental restoration and for finding more diverse and sustainable uses of the forest.
- The recuperation of water sources, forests and agricultural systems, with the condition that these satisfy the basic needs of the poorest sections within the communities.
- Long-term viability of restoration efforts, notwithstanding adverse macro-economic forces and government policies that obstruct real participation.
- Neutralising the despair and self-devaluation within communities, and promotion of a positive attitude.

The Situation in the Central Pacific

In the mountain area of the Central Pacific, with its semi-humid and humid tropical forests, the production and conservation of water depends on a combination of factors. Vegetation reduces the direct impact of the tropical rains. The porous volcanic soil acts as a sponge with cavities within the mountains functioning as water reserves. The whole system acts as a large regulator of the water flow in the rivers. The vegetation that collects rainwater and regulates its release consists of a particular type of flora which is characterised by an abundance of epiphytes. Concentrations of this vegetative type are found coexisting with the major water sources. Furthermore, regular wind convergence in particular geomorphological positions releases cloud humidity in these regions. In other words, there exists a complex relationship between the microclimate and the vegetation.

The botanical composition of these water-collecting systems is very diverse and presents many degrees of interdependence, generated by the co-evolution of the species. Such composition varies according to the altitude and the amount and type of rain. Hence, there is a correlation between the distribution of rain during the year and the type of vegetation that establishes itself locally. Under pristine conditions, the forest consists of three to five photosynthetic layers and a complex epiphytic presence in the form of bromelias, ferns, mosses, lichen, orchids and lianas. Disruptive intervention into such a complex system creates enormous damage with many unknown consequences, such as a sudden dramatic increase in local transpiration into the atmosphere.

Causes of Decline

Between 1960 and 1980 the government actively promoted agricultural expansion in the Central Pacific through slash-and-burn forest clearance. In particular, the aim was to create pastures for cattle in order to boost the export of meat. Consequently, this complex five-layered forest ecosystem, which contained thousands of plant and animal species, was reduced to a single-layered system with just a few species left. The majority of water-sheds in this region were degraded and the water harvesting capacity seriously undermined.

Once these pastures were created, the regrowth of endemic pioneering plants had to be controlled by machete and dry-season burning. The government then encouraged agricultural modernisation with the introduc-tion of agrochemicals, and 2.4 D-type herbicides were used to suppress native vegetation in the new pastures. The constant use of these herbicides encouraged the dominance of graminaceous plants (*Hypharrheniam*, *Imperata* and *Paspalum*). These slender-leaved species caused a rapid disappearance of broad-leaved plants that constitute the majority of the native flora. Only occasionally would *Sida*, *Lantana*, *Asclepias* and *Pteridium* ferns appear.

This reduction in diversity led to a chain of degradation. Diminished soil cover and reduction in the amount of organic material made the soil more susceptible to erosion and compacting. The continuous use of herbicides, erosion and compacting led to a decline in land productivity. The ecosystem rapidly lost the capacity to regulate the water flow, further accelerating erosion. This, in turn, had a direct impact on the livelihood of local com-munities. Even on the coast, traditional fishing communities suffered, since sedimentation destroys the coastal marine ecosystem.

Already by the mid-1980s the deterioration of the water sources had caused a progressive scarcity of drinking water in large areas in the Central Pacific. Moreover, the quality of the water had dropped in almost all the sources. The same climate changes further increased the desiccation of water sources. During the wet season large mudslides caused inundations in the areas downstream. The government's response was to favour the monoculture of imported species. But since these trees do not allow the growth of other species in their vicinity, they are unable to protect the soil from intense rain showers.

The Process of Natural Recovery

The forces of regeneration in plant cultures native to the area, if left to act on their own, will establish a pioneer vegetation. Later these pioneers will allow other types of plants which would progressively regenerate the diversity that originally characterised the area. This process of natural recovery through a succession of vegetation involves, in the beginning, small herbage and ferns; afterwards, bushes; next, softwood trees tolerant of light; and, finally, an appropriate microclimate which makes it possible for plants of the original understorey to settle in. This system will sooner or later come to resemble closely the physiognomy of the primary (original) forest. Before the massive destruction of the forests to establish agro-exports, the process of vegetal succession was based on an abundance and diversity of sources of seeds. To a great extent, also, local human cultures had adapted to the ecosystem.

Two decisive factors helped to start the recuperation of the water sources. First, it was important that the local communities were prepared to allow for a process of semi-natural recuperation in the affected areas. Second, it required the identification of native flora and fauna which originally formed part of the water sources. Because of the progressive degradation of large areas in the region, the availability of seeds – both in quantity and quality – is very limited. Thus the farmers of Arbofilia started to collect and reintroduce seeds of pioneer herbage and trees, without, however, removing what was left of the native vegetation. By observing the schedule of succession of vegetation they witnessed the sequence of some of the native species that have been reintroduced to rebuild the system. Bit by bit, farmers observed the vigorous, dense and protective covering of the soil returning. As the water sources recovered and stabilised, similar methods of regeneration could be applied to agricultural lands. In addition to mimicking the natural forest, this system is free of dependence on government assistance and on agrochemicals used in the monocultural plantations

Analogue Forestry

When Arbofilia started designing productive systems and strategies for environmental restoration it could benefit tremendously from experiences with the 'analogue forestry' approach. Analogue forestry is a system of silviculture that seeks to establish a tree-dominated ecosystem analogous in architectural structure and ecological function to the original climax or sub-climax vegetation community. There is an emphasis on trees that yield valuable products. Analogue forestry helps to empower rural communities both socially and economically through the use of species that provide marketable products. The system has applications to the sustainable production of food and other products, yet it always entails the highest possible proportion of local biodiversity. It also helped demonstrate that if all crops in the new forest are grown organically, many species of mammals, birds and flora confined to small pockets of rainforest can re-establish their populations within new micro-habitats provided by the analogue forest.

Attempts to move agro-ecosystems towards an analogue forest have been made in different parts of the world, but it was in Sri Lanka that formal experiments with analogue forestry began. This pioneering work was undertaken by the Neo Synthesis Research Centre (NSRC). The Environment Liaison Centre International (ELCI), based in Nairobi, established the Analog Forestry Network, with Arbofilia as one of the founding members.

The application of analogue forestry requires that local families take responsibility for their landscape and culture. The incorporation of the traditional system of bean cultivation (*Phaseolus vulgaris*) encouraged families to achieve these two objectives. People occasionally clear forest vegetation and sow the beans on a rotational basis, thereby allowing the natural vegetation and soil to recover. The products from analogue forestry – such as fruit, nuts, honey, beans, fibres, medicines or cacao – help to sustain the farmers' economy. Analogue forestry helps to minimise dependence on one single crop and helps farmers to diversify their family enterprises. The benefits which are being generated, however, are not necessarily monetary ones. Instead, we would prefer to cite the principle of 'the dignity of the people and the land'. This approach thus contrasts with conventional development projects that have as their scope 'to increase income', thereby transforming marginalised people into consumers.

Obstacles

When we designed and planned our restoration strategy, we had to take into account a number of social and political realities. Because forest policies are predominantly designed to favour interests and interventions foreign to local cultures, they deny these cultures real participation and security. Development agencies and private foundations make funds available for 'environmental restoration'. But these are in fact used to cover the land in monocultures, with the active support of government. The state, wishing to hold on to its privileges, assumes tasks and responsibilities which by their very nature should rest with the local communities. It thereby fails to facilitate local forest management. Moreover, the government's key tasks of auditing and controlling are poorly performed and corruption is deeply rooted within the forestry sector. It will take the commitment of several generations to resolve this problem.

Another obstacle is ownership patterns in the Central Pacific, which are changing dramatically. The government has encouraged foreigners to purchase land through fiscal incentives and other benefits. Local landowners and absentee landlords sell their land, thereby leaving tenants and sharecroppers with no option but to migrate, notably to the cities (a process hastened by poor services in basic fields such as education and health care, and a rural infrastructure which privileges the agro-export zones). Inadequate government control allows profits from drug trafficking, in some cases, to be invested in such land purchases, offering the Mafia a convenient foothold.

A further constraint is that the conceptual basis of the forest law is very weak. Although consultations were conducted prior to its enactment, none of the recommendations suggested by environmental and indigenous organisations were included. As a result, the law does not acknowledge that the highest degree of biodiversity is to be found in the non-arboreal components of the forest.

Probably the greatest obstacle is the absence of democratic mechanisms. Decision making is the prerogative of elites. There are hardly any consultative mechanisms which allow ordinary people to be involved in policy making and implementation. The democratic powers of the people are limited to voting for one of the leading families during general elections every four years. Government employees in general show little regard for rural people. Letters to officials, for example, are often left unanswered. A general feeling of apathy, if not despair, evolves among rural people as they notice that economic interventions in their region are not meant to improve their welfare. They experience imprisonment within a role that is merely instrumental – as wage labourers, for example. This undermines the confidence of rural families and dampens their motivation to engage in local ecological, cultural and economic restoration.

The process of economic and political globalisation is not altogether harmful. The problem is that it unleashes unprecedented deregulation and facilitates short-term exploitation of natural resources to the detriment of local people and the ecology. Moreover, it leads to the total monetarisation of relations between people and their relationship with the natural environment; and an enormous concentration of wealth and power. These are some of the reasons why globalisation threatens efforts at restoration and the improved management of natural resources.

In addition to this host of external obstacles, in our work we came up against a range of local problems which entrenched the process of ecological and cultural erosion and hampered Arbofila's efforts to mobilise the population in restoration campaigns. The Central Pacific has lost its indigenous cultures. Local traditional and spiritual leadership is absent. One has to cover large distances to consult traditional wisdom in support of local environmental and cultural restoration. A great majority of the present population did not originally belong to the area and do not readily identify themselves with the destiny of their landscape. The media, meanwhile, have a great impact on the rural people, promoting a consumerism that undermines sustainable practices vis-à-vis the natural environment. Most people in the area receive only very limited formal and informal education and the level of environmental awareness is very low. Consequently, rivers are polluted with domestic waste and illegal activities such as poaching are well supported. It is difficult for environmental groups to find acceptance and support for their work. Since most people live at a subsistence level, most of their time and energy is required simply to earn enough wages and to produce enough food. They are often reluctant to engage in endeavours other than those which help to satisfy their basic needs.

Last reflections

The process of environmental, cultural and economic restoration by the communities with whom Arbofilia is associated is above all an example of working with dignity for the land and the people. As the restoration – the strengthening of the 'alliance of life with life' – moves forward, the economic opportunities are becoming stronger, culminating in increased sustainable economic space. What has been achieved so far is mainly the result of positive work at the field level. The recuperation of the forest and the water sources, and the increase in economic benefits, are directly related to the level of local biodiversity. Farmers and indigenous people have planted more than 400 species of native trees and more than 200 species of fruit trees. In the Central Pacific and a selection of other regions approximately half a million trees were planted. Through local collaboration more than 70 micro-basins for the production of drinking water have been restored.

These results may look insignificant if compared with the range of destruction, but they give moral and technical authority to propose a strategy for restoration. Analogue forestry has proved a helpful component of the strategy. It has facilitated the rapid transformation of monocultures into polycultures, recuperated impoverished lands and restored water sources. New habitats and shelter have increased local biodiversity. The effects of climatic changes on agricultural and horticultural production have been mitigated. Moreover, local families have been mobilised in support of the ecological restoration process. This has created additional tangible economic and cultural benefits such as a more diversified range of products for domestic use and for the market.

BIBLIOGRAPHY

Boza, M. and Mendoza, R. (1980) *Los Parques nacionales de Costa Rica*, INCAFO.

Emmons, L. H. (1990) *Neotropical Rainforest Mammals: a Field Guide*, University of Chicago Press, Chicago and London.

Hartshorn, G. S. (1983) 'Plants: Introduction' in D. H. Janzen, (ed.), *Costa Rican Natural History*, University of Chicago Press, Chicago and London, pp. 118–57.

Herrera, W. and Gomez, L. K. D. (1993) *Mapa de Unidedes Bioticas de Costa Rica*, US Fish and Wildlife Service, TNC, INCAFO, INBIO, Fundacion Gomez-Ducunas.

Holdridge, L. R. (1967) *Life Zone Ecology*, Tropical Science Center, San Jose.

Janzen, D. H. (1983) *Costa Rican Natural History*. University of Chicago Press, Chicago and London.

Rachowiecki, R. (1991) *Costa Rica*, Lonely Planet, Hawthorn, Australia.

Savage, J. M. and Villa, J. (1986) *Introduction to the Herpetofauna of Costa Rica*, Society for the Study of Amphibians and Reptiles.

Stiles, F. G. and Skutch, A. F. (1989) *A Guide to the Birds of Costa Rica*, Comstock, Ithaca, NY.

9 Between Logging and Conservation
Traditional Management Practices in the Cameroon Rainforest

CENTRE POUR L'ENVIRONNEMENT ET LE DÉVELOPPEMENT

(CED)

Cameroon

Ecosystem and People

Bordered to the north by the Lobe River, to the south and east by the Ntem River, and to the west by the Atlantic Ocean, the Campo region in the southwest of Cameroon is famous for its paradoxes. The area was raised to the status of game reserve by the French colonial administration in 1932. Renowned for its immense biological diversity, particularly on Dipikas Island, the region was selected as the site of a Global Environmental Facility (GEF) biodiversity conservation project. Yet since the beginning of 1996 the same area has become host to La Forestière de Campo, one of the biggest logging companies in Cameroon, a subsidiary of a French holding company. The 6,000 local inhabitants – many of whom engage in subsistence hunting and farming – are subject to the rules of the reserve, whereas these restrictions do not seem to apply to the company.

Of the approximately 300,000 hectares covered by the reserve, only a relatively small area – 36,000 hectares – is inhabited by people. This is a heterogeneous population composed of the following groups: 300 Bakola Pygmies who live in scattered groups; the Mvaes, the most populous tribe in the region; the Batangas and Mabeas, who inhabit the northwest of the reserve; and the Yassas, fishermen and farmers, who live along the coast. These people live by cash-crop and swidden farming, fishing in the Atlantic and in small rivers, and by hunting.

The Dja-et-Lobo region covers about 19,804 square kilometres. This area is inhabited by 134,916 people of several ethnic groups: there are the Fang, Bulu, and Fong; some small groups like Mvem, Zaman and Makia; and some 5,000 Baka Pygmies, who practise agriculture, fishing and hunting, and harvesting of non-timber forest products. In spite of the significant population and their dependence on the natural environment, all forms of exploitation are prohibited by law in the Dja Reserve, which was created in

1950. The area was given the status of a UNESCO Human Heritage Site in 1987.

The region's forests are rich in both flora and fauna. Timber species of great commercial value, such as African teak (*Chlorofora excelsa*), are abundant in the forest. This accounts for the presence of some ten logging companies in the region. The fauna include large mammals such as the forest elephant of the rainforest belt (*Loxodonta africana cyclotis*), the African buffalo (*Syncerus caffer nanus*), gorillas (*Gorilla gorilla gorilla*) and panthers (*Panthera pardus*) – although their numbers are declining under pressure from intensive hunting. There are also more than 200 species of medium-sized and small mammals. Some 100 species of birds have been identified, of which the eagle is the largest. As for fish, 191 species have been identified in the Sanaga River basin in the north, and about 149 in the Congo River basin in the southeast.

Local Forest Use and Management

Two modern laws determine the state's relationship with the forest: Ordinance No. 74-1 (6 July 1974), which governs ownership of private and state land, and the Forestry, Wildlife and Fisheries Law (9 January 1994). As a general rule, all forest land is owned by the state or local councils, whereas customary law clearly recognises local peoples' ownership of land. In this clash between modern and customary law, the state continues to show its bias towards modern law. This means that the state can authorise the industrial exploitation of forest by outsiders. Although local people lack formal rights over the forest, they can negotiate to some extent with the logging companies for compensation. In the past such agreements, so-called *tenue de palabres*, were endorsed by the Cameroonian government. This helped to mitigate conflicts and introduced at least some benefits – like access roads, schools and playgrounds – to the villages. With the enactment of the 1994 law, however, the system of *tenue de palabres* has been repealed.

In spite of the absence of official recognition, local people continue to organise resource use according to well-defined customary laws, based on a vast reservoir of traditional knowledge. The Baka and Bakola people derive most, if not all, of their basic requirements from the forest. To build the *moungouloum*, or traditional hut, they need small trees or branches and *irage* leaves. For the *barty*-style houses, they use the trunks and leaves of bamboo for the walls and roof. Beds and benches are also made of bamboo. They cut cane to make ropes, used in house building and furniture making.

Game in these areas is unequally distributed. While animals are plentiful in the Campo Reserve and the immediate vicinity of the Dja Reserve, game is less abundant in other parts of the Dja region. There, the most common species are ordinary hedgehogs (*Thryonomidal*), rats (*Cricedidae*), civets (*Viverridae*) and buffaloes (*Bovidae*). Hunting is carried out either for subsistence purposes by local peasant farmers or for commercial purposes – mostly by professional hunters who supply the large urban centres. Local people generally complain about competition from professional hunters who command considerable financial resources and often use modern equipment. Commercial hunters sometimes benefit from lax law enforcement by government personnel responsible for wildlife management. Local hunting is usually carried out with traps, though bows and arrows and rifles are also used. There are no reliable statistics on quantities of wildlife taken out of the region, although local people express their dismay over the disappearance of several species.

Table 9.1 Extinct or very scarce species in Mfouladja and Campo areas

Local or common name	Scientific name	Extinct	Very scarce
Okweng	*Bovidae* (*cephalophe de Grim*)	X	
Nkok	*Bovidae* (*Guib harmache*)	X	
Zip	*Bovidea* (*Cephalophe a dos jaune*)	X	
Mvin	*Bovidae* (*Cephalophe de Peters*)	X	
Seuk	*Cercopitheciae* (*Mandrill*)		
Buffalo			X
Pangolin Geant			X
Panther			X

The Baka and Bakola hunt, fish and collect forest products such as wild yams, honey, wild berries, palm oil, *moabi* nuts and *ilomba* seeds. Their traditional medicine is essentially based on trees and herbs. For safety reasons, trees are used as an encampment when the night is spent in the forest. The bark of the *bologa* tree, or *gaka* in the Baka language, is collected and burnt, as its smell keeps wild animals away and reportedly makes leopards sneeze! Many forest products are used for making tools. Mortar and pestle are made from wood. Tree bark is used as a grinding stone, while the kneecap of an elephant serves as a millstone. Plates, spoons and trays are made from wood. Oyster shells are used as scrapers. Baskets, sacks and mats are made from cane and reed. Grated or burnt wood, for example from the *baa* tree, gives a red powder, and oils are used for cosmetic purposes. Some wood carving takes place. The mango (*irvingia*), *bubinga* (*Guibourtia demeusei* and some other species), and *moabi* (*Baillonella toxisperma*) trees play an important role in the lives of Baka and Bakola people. Another tree, locally called *kaki* (*Diospyros kaki*), is believed to produce thunder.

Firewood, still the main source of energy in the area, is collected in the form of dead branches and tree trunks. On the average, over 50 per cent of all households in the area consume ten kilogrammes of firewood per day. In those areas where there is an industrial sawmill, conflicts sometimes arise when local people are denied the use of wood residue. There have even been cases where a forestry company burned its wood waste to prevent local people from using it for charcoal. In some areas, particularly on the outskirts of large towns, the exploitation of firewood poses a direct threat to the forest. In the Mfou area, for example, commercial exploitation of wood has considerably degraded the forest. Ever since the devaluation of the CFA franc in January 1994 and the subsequent increase in prices of cooking gas and kerosene used by urban populations, firewood consumption in Yaounde has risen. In the Mfou-Awae region, the main area supplying the capital city, there is an urgent need to find solutions to this problem which also offer alternatives to small-scale firewood dealers.

Occasionally, there is fierce competition between villagers and logging companies for certain commercial species that are used locally. Several cases have been recorded where trees set aside by families to be used for building projects have been cut by logging companies. Often no financial compensation is offered by the company, under the pretext that people lack legal title to the land and that the trees were inside the company's logging concession. There are also frequent conflicts over the *moabi* tree (*Baillonella toxisperma*), whose bark is used for medicinal purposes and whose fruits provide oil for local consumption, and over the *essingan*, said to be a sacred tree that protects people from evil spirits. Both these species are among the most commercially valuable timber species, and are logged for the export market.

It is interesting to note that there are almost no conflicts related to the use of these trees among the communities themselves. There are, for example, widely known and accepted rules governing the use of *moabi*. If a villager finds a *moabi* tree in the neighbourhood of the village, he or she must mark the tree by cutting the grass under the tree, and inform other villagers about its location to prove ownership. Having done so, the family has a dominant user right over the *moabi* fruits. *Moabi* trees located in the vicinity of the village, however, are considered to be the common property of the village.

Pygmy communities maintain a strong attachment to their traditions. These include a great respect and awe for the *Djengui*, or forest deity, who is believed to protect the Pygmies. The relationship with the forest is determined by the Pygmy perception of the forest as the home of the *Djengui*, and hence the meeting point between the spirit world and the Pygmies. Since the Bantus in this area are almost all Christians, there are very few examples of Bantus using the forest for religious purposes. Nevertheless, the forest is filled with places and species whose reputation elicits fear and respect from Pygmies and Bantus, as they are perceived to possess supernatural powers. Lobe Falls in the north of the Campo region and the shifting sands in the Alati region are two places believed to be inhabited by spirits, good and evil.

Table 9.2 Some reputations of certain animal or plant species that have proven to be protective

Local name	Common name	Scientific name	Reputation
Oveng		Guibourtia tessmanii	Its bark is reputed to expel evil spirits. Some pacts concluded in the society are connected to this tree.
Doum	Cheesemonger	Ceiba pentandra	This tree is reputed to shelter witch meetings at night.
Wo'o	Chimpanzee	Pan troglodytes	Totem animal very famous in the 'practice' of a local rite called So.
Zeu	Panther	Panthera pardus	Totem animal considered as a protective force for witches and wizards.
Yiii	Cobra	Naja nicricolis	Totem snake reputed to foretell calamity or unfortunate events.
Mvom	Boa		Totem snake. Its presence in a river limits all exploitative activities.
Akoung	Owl		This bird is not eaten because it is reputed to animate meetings.
Engbuwang	Crow		This bird is not eaten or shot because it is reputed to eat human flesh.

Today, farming in the forest zone is practised with rudimentary tools and few modern inputs. Some 60 per cent of peasant farmers cultivate an area smaller than one hectare. This finding goes against the common belief that smallholders constitute the principal threat to the forest in Cameroon. It must be mentioned that the proportion of forest converted into farmland is substantial as swidden farming requires long fallow periods and, hence, a large land base. The fallow periods do, however, allow for undisturbed forest regeneration. On the other hand, new pioneers who grab primary forest land for industrial plantations constitute a serious threat to the Pygmy communities. These are concentrated to the east of Mintomb where vast areas are cleared each year by rich white-collar farmers.

Traditional rules govern the use of forest and land by local communities. Although these customary laws vary from region to region, there is a common understanding about resource ownership. Common to all these laws is the provision that land, including the forest covering the land, belongs to the village. Boundaries are known to local inhabitants and to neighbouring villagers. A Baka villager, for example, is able to draw the map of his village's communal land in detail and to the right scale. An exception is swidden farm plots inside primary forest, which become the private property of the farmer even if the land is left fallow. Private ownership can also be obtained through inheritance or donation.

External Influences

Communities in the Campo and Dja regions have been managing their forests sustainably without outside interference for generations, using a combination of swidden farming, hunting and gathering. Under French rule, a process of resettlement along the main access roads took place. This was in line with the designs of the colonial regime, which aimed to increase control over the local population. This has been compounded by the intervention of new, mostly foreign, actors in the management of forests. These include the state, foreign logging companies, researchers, local entrepreneurs, bilateral and multilateral development agencies, conservation organisations, NGOs and foreign research institutes. They are introducing new ways of thinking, more often than not in conflict with the traditional long-term management logic of the local communities. The World Bank, which has attempted to introduce greater 'transparency' and 'efficiency' in forest management, has been one of the most powerful actors since the beginning of the 1990s. The Bank's interference in the forestry sector is directly related to its structural adjustment programme for Cameroon.

The depletion of forests in major timber-exporting countries, such as Côte d'Ivoire and Ghana, led to a shift by logging companies to Central Africa, which harbours the last extensive tract of African forest. The devaluation of the CFA franc in 1994 intensified this shift, as the Cameroonian government sees operations by foreign logging companies as a major source of hard currency revenue. Companies operating in the Campo and Dja region are the Dutch firm Wijma and the French firm La Forestière de Campo.

In addition to loggers and bankers, Cameroon has attracted the attention of foreign conservation organisations, donor agencies and research institutes, such as the Dutch research institute Tropenbos, the GEF,

ECOFAC (Le Programme de Conservation et d'Utilisation Rational des Ecosystèmes d'Afrique Centrale), the European Union, and the Dutch Department for International Cooperation (DGIS).

As a result, the Cameroonian government now faces claims to forest from all directions. Commercial loggers, donors and conservation agencies offer the government money, and political elites profit from these contacts. Notably prior to elections, one can witness a steady increase in the number of concessions handed out to logging firms. But local communities tend not to generate revenues for the state, nor do they mobilise political pressure. The government thus engages exclusively in agreements with external parties, dividing the forest into 'production forest' and 'protected forest'. Local communities are crushed between these forces. This is the prevailing situation in eastern Cameroon.

Both logging and conservation activities are large-scale and conducted under time pressure, allowing for little or no liaison with local communities. Logging and poaching are short-term, with little or no regard for conservation, and no interest in sharing management with local people. The conservation logic, according to which whole areas are turned into sanctuaries, still suffers from a general lack of consideration for the interests of people inhabiting the area. Conservation projects generally claim that they are working for the benefit of mankind and future generations. But they do so at the expense of the local population, which is apparently not looked upon as belonging to 'mankind'.

In this respect, the GEF/DGIS-sponsored conservation project in the Campo Reserve may offer an important opportunity to reconcile social and conservation interests. The outcome will depend on the approach adopted by the funding agencies. The project aims to further develop and consolidate a core zone and buffer zone. Both Baka and Bantu communities have been living in this region for generations. A number of communities are now at risk of being evicted from the core zone. It is strongly recommended that DGIS and GEF permit these communities to manage community forests. Such an approach would represent a major breakthrough, as it would offer local people a meaningful way to become engaged in the management of the reserve. Much of the social hardship could be avoided.

Communities try to adjust to the changing circumstances as best they can. The Bakola communities of Campo Nazareth and the Bakas of Djoum Subdivision have for many years been leading a settled life, although they still depend on hunting for their subsistence. Traditionally, they hunted everyday, but intensive poaching has made game scarce in the immediate vicinity of the villages. Thus they are forced to undertake long hunting expeditions deep into the forest. They then smoke their catch, which may last for several months.

State policy is encouraging forestry exploitation. During a press conference in August 1995, the Minister of Environment and Forestry declared that Cameroon's forests were under-exploited and that over the next century five million cubic metres could be extracted annually without altering the national forestry potential. The revocation of the *tenue de palabres* system in 1994 reflects how the state has sided with logging companies at the expense of local people. Perceptions of the impacts of logging vary according to the different groups. Members of the local elite – notably wealthy traders, entrepreneurs and retired government officials – tend to emphasise the benefits of logging. They have enjoyed substantial advantages with the arrival of the logging companies, such as free timber, fuelwood, second-hand cars, and jobs. They are the companies' local allies and tend to highlight the benefits of logging for the villages, in the form of roads, health facilities and schools, and to present the negative impacts as the 'normal costs of development'. Local people who are employed by the logging firms, and others who also depend on these companies financially, suffer from poor and unsafe working conditions, but they prefer to take a neutral stand.

The great majority of villagers, meanwhile, seem opposed to logging, or at least resentful of the way logging is currently being practised. Their main complaints are clear enough. Few social or financial benefits accrue to the village, whereas villagers will face the consequences of forest degradation after the departure of the logging companies. Moreover, the companies do not comply with regulations and agreements, for example, in denying villagers access to residual timber from the sawmills. The companies cause social destabilisation by buying the attentions of villagers' daughters and wives. This has in many cases led to broken marriages and the destruction of young people's future prospects. The Bakas and the Bakolas indicate that an increasing number of people now suffer from sexually transmitted diseases, problems they never encountered before. Game is becoming scarce, dispersed by the noise from logging operations, and poaching by both employees of logging firms and professional hunters has intensified. These hunters use the access roads to get to the interior of the forest, and logging tracks to smuggle out bushmeat. A number of much appreciated timber and non-timber products – used by villagers for medicinal purposes, as food or as building material – have become scarce or have disappeared. Even villagers' farms are being destroyed by road construction or logging operations. Farmers do not receive compensation when this occurs, as the firms maintain that the villagers have no legal title over the land.

The 1994 forest law introduced the concept of 'community forests' as a consequence of pressure on the Cameroonian government by the World Bank. The idea of granting local communities ownership over forest land was unacceptable to the government, however, and the forest law therefore states clearly that community forests are state property. Local communities are permitted only user rights (*droits d'usage*) under the strict supervision of the forestry administration. Community forests, furthermore, are granted only to communities that make a formal request. As a result, there are no 'community forests' in the Campo and Dja regions. Local people are not aware of the option, and the forestry department has been unenthusiastic about implementing the policy. One important action to be carried out in this area would thus be to popularise the provisions regulating community forests, and to enhance the capacity of the local population.

Local Initiatives, Problems and Needs

In the Campo and Dja regions, traditional forest management is confronting a range of obstacles. Within the communities, social control mechanisms are quite weak. Part of the reason for this is that the social structure is highly egalitarian. Inhabitants of the forest of south and east Cameroon belong to the 'acephalous' social groups, characterised by horizontal organisation and the absence of a single decision-making authority with full coercive powers. This hinders the imposition of rules to regulate the management and use of forest resources. In southern Cameroon, because both Baka/Bakola and Bantu communities lack strong chiefs, there is no authority to voice local peoples' claims over forest land.

Cultural restrictions that have some degree of success are food taboos and mystical beliefs. For example, the Baka Pygmies do not eat chimpanzee or gorilla meat, and the *essingan* tree is protected for its mystical powers. Sacred forests are small tracts which are believed to be the dwelling place of the totem of the King and eminent men. Nobody, including forestry department personnel, is allowed to enter the forest without the prior consent of village authorities, or to cut trees or clear farmland. Probably as a result of the influence of Christian beliefs, however, no such forest sanctuaries exist in the central and eastern provinces of Cameroon, although they can still be found in the west of the country. It is important to note that forest protection is not considered to be a goal in itself by local communities. The most important motive behind these 'conservation' activities is the compliance with customary practices and local taboos, in order to gain the favour of the ancestors who are believed to have the power of providing resources for the well-being of the living.

Local forest management is severely hampered by a lack of awareness among local people. Although many adhere to customary laws, these are being undermined by modern land and forest laws about which local people have no knowledge. Nevertheless, there are many aspects of the 1994 forestry law that could be used by local communities: the definition of the forest-related rights and obligations of local communities; procedures for establishing community forests; and the rights and obligations of local people *vis-à-vis* logging firms and plantation owners.

In reality, villagers are not in a position to check outsider exploitation of species that are either protected or of great importance for their own livelihood. They perceive the forestry administration as an institution whose job it is to curtail local peoples' use of the forest, instead of as a partner offering guidance and support in forest management. Lacking information about protected flora and fauna and poaching prohibitions, and without backing from government, the community is unable to contain the poachers. In one case villagers living close to the Dja Reserve managed to capture a gang of poachers carrying modern rifles who entered the village territory. The captives were taken to the local government headquarters. To the villagers' great dismay, the poachers were handed back their arms and set free, while the villagers were arrested and kept in jail for a number of days.

The standard of environmental awareness is low and little is done to change this. Food taboos do allow for the protection of some flora and fauna. But people still regard the forest primarily as a source of agricultural land. This view is shared by more than 70 per cent of the inhabitants of Djoum, Campo, Mfouladja and Zoétélé regions. Ecological functions of the forest – as a stabiliser of the microclimate or as a source of fauna and flora – do not receive priority. It is a further obstacle that where Bantus and Bakas live close together, as is often the case, tensions arise that make it difficult to engage both groups in the establishment of community forests.

External obstacles are also undermining local forest management. The 1994 law creates much uncertainty. Not only are community forests defined as state property, but the law fails to define 'community': it is unclear whether the term applies to a village, a clan or a group of villages. Conflicts between neighbouring 'communities' are thus likely to arise. This is all the more serious as the law indicates that, if conflicts occur, the forestry administration has the right to withdraw the licence.

Industrial plantations and forestry companies are another serious impediment to local control over forests. This is most evident in the Mintomb area in the south of Djoum, where provincial elites are establishing plantations of export crops and competing with the local population for land. The operations of La Forestière de Campo have led to the destruction of forest plots that were being managed for the extraction of essences, and have opened up the forest with roads that are also being used by poachers.

Recommendations

Given the difficulties faced by local people to maintain and develop traditional forest management practices, we propose the following measures:

- It is important to improve the possibility of establishing viable community forests. Local people must be informed about the potential of community forestry and the relevant section in the new forestry law. As a degree of locally institutionalised control over forest use is indispensable, more discussion within the communities on these issues is needed. It would also be advisable to appoint a contact person from within the community who could liaise with the forestry department. Finally, communities should start the process of submitting formal applications to the department for community forestry licences.

- Measures need to be taken to reduce the pressures on wildlife. We suggest that local communities should be encouraged to help control the poaching. In some communities, notably among the Pygmies, a number of taboos that prohibit or regulate the hunting of certain animals still exists. In some Bantu areas, traditional hunting is less intensive during the animals' mating season. It is important to recognise the role of these taboos and to bring back traditional practices that enhance respect for the environment. In this way, local people could be involved directly in wildlife protection. In addition, programmes for breeding small game would increase local income generation, and might also reduce hunting pressure on these species.

- Shifting cultivation, the predominant farming system practised in the area, is a major cause of forest degradation. Intensification of farming to diminish the area used is therefore essential. Experiments with biological agriculture must be supported, and extension of such practices should be promoted. Only with such measures can land degradation from fire and soil erosion be reduced.

- Communities require training on land rights and forestry law, emphasising the rights and duties of local communities *vis-à-vis* the state, logging companies and industrial plantations. As conflict grows around forest management, it will be equally important to strengthen local communities in conflict management. Efforts should be made to use the legal concept of community forestry to secure land more or less exclusively for Pygmy communities, so as to lessen the influence of the neighbouring Bantu population.

- Attention needs to be given to social activities that enhance group cohesion. In this respect, it is important to promote small-scale appropriate technology that enables sustainable harvesting and processing of non-timber and farm products. For this to work, it must generate income and improve local skills. In their conservation programmes, DGIS and GEF should make community forest management within the project area a condition of their funding in the Campo Reserve, and see to it that established communities in the core zone are not forcibly resettled.

Recognition of traditional knowledge is a precondition for these measures to be effective. Detailed inventories of such knowledge must be compiled, and shared with local people so that it does not get lost. This should be done in a participatory manner. In addition, external funds must be made available to help meet the costs of the proposed actions. Committed individuals working for NGOs and the government need to be prepared to work directly with forest communities. And to enhance the sense of security and facilitate long-term management, it is vital that both the forest law and forest policy of Cameroon be amended, enabling local communities to enjoy communal ownership rights over community forests.

10 The Challenge of Conservation in Kahuzi-Biega National Park

CENTRE D'ACTION POUR LE DÉVELOPPEMENT DURABLE ET
INTÉGRÉ DANS LES COMMUNAUTÉS (CADIC)

Democratic Republic of Congo

The Democratic Republic of Congo is Africa's third-largest country after Algeria and Sudan. Covering over 2,344,510 square kilometres, it extends from the savannah woodlands of the southern fringes of the Sahara, in the north, to the woodlands and grasslands of Zambia and Angola bordering the Kalahari in the south. It contains over half of the continent's tropical moist forests. The forested regions are so immense that it is still possible to fly for two hours over virtually undisturbed forest, from Bandundu in the west to Bukavu in the east. The Democratic Republic of Congo is renowned for its biological richness, with more species of birds and mammals than any other African country. This diversity is due both to the country's sheer size and to its range of geomorphological and climatic conditions. From west to east, it encompasses forests on the Atlantic coast and glaciers in the Rwenzori – at 5,119 metres, Margherita Peak is Africa's third highest mountain. The vast expanse of the equatorial forest lies in the Central Basin – the Cuvette Centrale – at an altitude of just 300 metres above sea level.

The Democratic Republic of Congo has some of the world's most extensive and spectacular national parks. Four of these are on the UNESCO World Heritage List for their outstanding biological value: Garamba, which consists of open grassland (920 square kilometres); Kahuzi-Biega, largely comprising forest (6,000 square kilometres); Salonga, which is entirely forest; and Virunga, which has extensive forest cover (7,800 square kilometres). Other parks include Kundelungu (7,600 square kilometres), Lomako, Maiko, Marungu Mountains and Okapi (36,000 square kilometres, including Salonga) and Upemba (11,730 square kilometres). The Democratic Republic of Congo's existing parks cover some 84,880 square kilometres, and several new protected areas have been proposed recently. Of these, the most significant area for forest conservation is the Ituri Forest, which has been created to protect the okapi and other species characteristic of this otherwise unprotected forest.

151

The forests are of great importance to local people for fuelwood, game, crops and other basic necessities, as well as for the production of export commodities such as timber and palm oil. The Democratic Republic of Congo has a population of almost 40 million people, 40 per cent of whom live in urban areas. The cities are located around the periphery of the Central Basin. The economy of the country is based on a mixture of mining and agricultural activities. Copper, cobalt, diamonds, crude oil and coffee are the main export products. There is some industry, such as breweries, cement manufacturing, oil refining and hydroelectricity generation. The central forest areas have always been sparsely populated. In recent years, a fall in prices on the international market for the few agricultural commodities that the region produces has further weakened the economy of the area and, as a result, there is a gradual migration of the population towards the more prosperous areas to the south, west and east.

Once known as the Belgian Congo and then as Zaïre, the Democratic Republic of Congo gained its independence in 1960. A territory of great cultural diversity, the country has since then been ridden with conflicts inspired by the hunger for regional autonomy. This, together with an inadequate communication system, has seriously weakened the infrastructure. Consequently, large areas are cut off from government control and many parts of the country suffer major depopulation. It is now more difficult to travel by river or road than it was 30 years ago and, in some remote parts of the east and north, the only postal, health and educational services available are those provided by missionary groups. Soon after this case study was written the country was plunged into civil war.

The history of the Democratic Republic of Congo would undoubtedly have been very different but for an accident of geography. The Zaïre (formerly Congo) River is navigable for most of its length and could have provided direct access for human penetration into the centre of the continent. But this route is broken by two sets of impassable rapids, the first below Kinshasa and the second below Kisangani. Railways now bypass both rapids, but the time and cost of shipping goods have always prevented exploitation and development of the interior. This fact has been of special significance for forest preservation. The cost of transporting timber from the forests of the basin to European markets has always been so high that only the most valuable species could be exploited profitably. This has meant that the central forest zone has experienced a low volume of very selective logging that has had little impact on the integrity of the forests.

Over 10,000 species of plants are known in the Democratic Republic of Congo, a number exceeded in Africa only by South Africa and, possibly, Madagascar. Of the 3,921 species described in the first ten volumes of the *Flora du Congo Belge et Ruanda-Burundi* (now published as the *Flora d'Afrique Centrale*), 1,280 were considered to be endemic. Many of these species will undoubtedly be found eventually in other countries, and there is now evidence that Gabon and Cameroon have higher floral diversity endemism. Nonetheless, floral diversity of the Democratic Republic of Congo remains exceptional.

The Democratic Republic of Congo also shelters 409 mammals, 1,086 species of birds, 80 species of amphibians (of which 51 are endemic) and 400 species of fish. Endemism occurs in all these groups, and many species or subspecies are considered to be in danger of extinction (IUCN 1990). The Democratic Republic of Congo has the highest diversity of primates in continental Africa, with 30 species, of which 19 species or subspecies are endemic. A notable example is the Pygmy Chimpanzee (*Pan paniscus*), one of the closest relatives to humans, occurring in forests in the Lomako and Salongo regions of the Central Basin. Other rarities include the Mountain Gorilla (*Gorilla gorilla beringei*), which inhabits the border areas of Ruanda, and the Lowland Gorilla (*Gorilla gorilla graueri*), which is found up to an altitude of 1,500 metres within the Kahuzi-Biega National Park. There are around 30 species of duiker, as well as the spectacular endemic okapi (*Okapi Johnstoni*), a forest giraffe. Some 18 species of fruit bats have been identified, and the mountain forests on the eastern borders of the country are particularly rich in birds. The forests on Itombwe Mountain are home to at least 564 species, the single richest area for birds in Africa. An estimated 25 bird species are endangered, of which eleven are endemic (Collar and Stuart 1986).

Although they are one of Africa's centres of floral and faunal diversity, the forests are being depleted rapidly and the Democratic Republic of Congo faces serious conservation problems. Deforestation is most severe in the provinces of Kivu, Shaba, Kasaï and Bas-Zaïre and around the capital, Kinshasa, where virtually all forest has been cleared within a radius of 60–100 kilometres. On the forest's periphery, especially in the Bas-Zaïre region, agriculture encroaches on forests after loggers have been through the area. Local people also cut wood for domestic needs. In Shaba and Kivu provinces, small-scale enterprises extract timber from the southern and eastern fringes of the forest, while the urban demand for charcoal and fuel-wood is on the rise.

As poverty spreads in the rural communities of the Democratic Republic of Congo, local people argue that the government and international nature conservation groups are more concerned about wildlife than they are about people's needs. Clearly, it is imperative to take into account people's rights and aspirations in ensuring long-term protection of the ecosystem. We must find a balance between securing access to forest resources for forest-dwelling peoples and preventing over-exploitation.

Kahuzi-Biega National Park

The Kahuzi-Biega National Park is named after the two highest mountains that lie within its boundaries. It is located on the western slopes of the mountains of eastern Democratic Republic of Congo, southwest of Lake Kivu in South Kivu Province.[1] The park's altitude varies between 900 and (at the summit of Kahuzi) 3,300 metres. It consists of two major parts linked by a corridor: the smaller part (1,000 square kilometres), located northwest of the town of Bukavu and the main part (about 5,000 square kilometres), spreading south from the Bukavu–Irangi–Wlikale road towards Kisangani in Haut-Zaïre Province. The region has a mountainous climate with a dry season from June to August and a rainy season from September to May. The average rainfall is 1,900 millimetres and the monthly mean temperature is around 15°C.

The range in vegetation is due to a combination of differences in altitude and human interventions. Between 2,000 and 2,400 metres we find primary and secondary mountain forest (*Myrianthus holstii* and *Xymalos monospora*), and swamp forest (*Cyperus*). Between 2,350 and 2,600 metres, bamboo forest occurs (*Arundinaria sp.*). Beyond 2,600 metres are the dry savannah and grasslands.

Kahuzi-Biega was originally created for the purpose of saving the Eastern Lowland Gorillas, which were seriously menaced by poaching. Today this sub-species of gorilla survives only in eastern Democratic Republic of Congo. The population is estimated at between 3,000 and 5,000 individuals, of which 270 live in the hilly part of Kahuzi-Biega (Lee 1988, de Monza 1996). The gorillas live in families dominated by a fully grown male, called the 'silverback'. The largest known family group consists of 34 individuals. In Kahuzi-Biega, 25 groups and nine solitary males have been registered. Apart from these famous apes, other fauna living within the park include elephant, buffalo, chimpanzee, numerous species of other monkeys, antelope, giant forest hog, bushpig, bongo, as well as birds and reptiles.

The park came into being in 1970 as a result of a presidential decree. Initially, it covered only 600 square kilometres, but was extended to 6,000 square kilometres five years later. Kahuzi-Biega was included in the UNESCO World Heritage Site list in 1981. The park contains a natural corridor joining the original 600 square kilometre area to the later extension, descending from 2,000 metres to the forests in the lowlands. This corridor can be likened to an umbilical cord, which permits genetic exchange among animal populations in the plains and those in the mountains.

The People of Kahuzi-Biega

The original 600-square-kilometre park is one of Africa's most densely populated regions, with 300 inhabitants per square kilometre. Some 9,000 people from seven different ethnic groups – including the Rega, the Shi, the Tempo, the Kano and the Pygmies – live within the boundaries of the park.

The Rega, predominantly forest farmers, are the main group in the western part of the park. They practise shifting agriculture and hunting for subsistence, while gold-digging provides some cash income. The Rega's religious leader is the *Mwami*, their absolute ruler, who is initiated by a secret society. The *Mwami* controls land and social and economic life, as well as the entire forest, including the animals. Villagers require his permission to hunt, and must pay tribute if such permission is granted. The second-largest ethnic group is the Shi, whose political life is also determined by a *Mwami*. Like the Rega, the Shi are primarily agriculturalists.

There are currently three types of land tenure in the area. *Bwasa* is the system whereby the *Mwami* permits villagers to cultivate pieces of land for a specified period of time. According to the *kalinzi* tenurial arrangement, the *Mwami* – who by tradition owns the land – sells land to villagers, after which it cannot be expropriated. From the *kalinzi* system evolved the *Bugule* system. Under this arrangement, the new owners still depend on the *Mwami* whose permission they require to actually use the land.

The portion of the park that stretches out over high altitudes is suitable for large plantations (quinine, tea, coffee) and cattle grazing. Here the Shi is the dominant group. Their presence has led to an accelerated replacement of the forest by agriculture and pastures. Attracted by virgin land with high agricultural potential, Shi villagers here prefer to convert the forest into farmland, with a few permanent farms now interrupting the continuity of the Kahuzi-Biega. This creates other pressures on the park. Due to the disappearance of forest in the *Bushi* (Shi territory), villagers try to satisfy the increased demand for wood and construction timber by taking wood from the park. Many young people engage seasonally in intensive gold digging along the rivers. To provide themselves with food, these young people enter the park to trap monkeys. The division of labour is such that women concentrate on the production of food, men on hunting and young people on the manual exploitation of gold. There is no division of labour on the basis of ethnicity or religion.

Another risk for the park comes from the rehabilitation and enlargement of the Bukavu–Kisangani road. This poses a real threat to animals in the oldest part of the park, and especially to the gorillas. Traffic frightens them and threatens to divide them into two small groups. If this occurs, the population will be doomed as a consequence of degeneration from inbreeding. The road also introduces other problems, such as the increase of poaching, road accidents involving gasoline trucks, and the danger of fires started by careless travellers.

A Conservation Project in the National Park

Following the creation of the new park boundaries in 1975, some 4,500 people (approximately 25 per cent of the total population living in or adjacent to the park) were resettled, and forced to start farming new land outside the park. These people continue to express the merits of their traditional conservation practices, which include regulation of hunting and access to land by the *Mwami*. It is, necessary to acknowledge, however, that such management practices sufficed in an era when the population density was relatively low and the fertility of the land surrounding the park was high. It is doubtful whether the traditional systems can cope with higher populations and lower soil fertility. The *Mwami* has indicated that he was never involved in the decision to extend the park boundaries to cover 6,000 square kilometres. He is dissatisfied with the current situation, as the park is an important source of food, medicine and fuel for himself and his people.

This was the situation when, in 1985, the Institut Zairois de la Conservation de la Nature (IZCN) and the German governmental aid agency Gesellschaft für Technische Zusammenarbeit (GTZ) started a project to enhance the protection of the national park – the Kahuzi-Biega National Park Pilot Project. Its purposes were to achieve more effective control over the park's territory, to create a better management structure, and to diminish the pressure of various local communities on the park by developing activities on its periphery. IZCN and GTZ aim to help people by establishing primary health care centres and schools, and by improving sanitary conditions and women's water supply projects. They also plan to support the introduction of agricultural techniques which are better suited to the region's soil and climatic conditions, and which generate an increased income for the inhabitants living in the environs of Kahuzi-Biega.

The IZCN started a national conservation strategy and awareness-raising campaign. It seems that these initiatives are bearing fruit, to the extent that the population accepts that the conservation of the park's ecosystem is also in their interest. The Germans contributed to that strategy by offering training courses for park staff and guards, and equipment to improve logistics and working conditions in the field.

The project's other objectives are to encourage scientific research and to regulate tourism. Park management wants to prevent the sort of tourism that could endanger the gorilla population. The impact of tourism on the livelihood of the local communities is most tangible in the oldest part of the park. Here local people make a living by selling handicrafts. This has increased their standard of living, notably in terms of housing and sanitary conditions. Yet tourism is the only means of commercial exploitation of the park permitted by law, and has also become an important source of income for the IZCN. This money should be invested in the development of the communities living on the periphery of the park. Unfortunately, the revenue is sent to Kinshasa.

The people who live around the park often approach IZCN/GTZ with requests for assistance to build schools, dispensaries and equipment to process agricultural produce, such as manioc mills, oil presses and roads. They argue that they lack sufficient financial means to realise these projects themselves. The staff of the IZCN/GTZ project argue, on the other hand, that the lack of cooperation and information are far more important obstacles than the lack of finances.

This pilot project is the only intervention by outsiders in the park. But the local population feels that the activities so far undertaken by the project are inadequate. As yet, these different measures, which are meant to diminish the local people's pressure on the park, have not succeeded in eradicating the clandestine clearing of forest. Staff from IZCN and GTZ estimate that cleared forest will take five centuries to regenerate. Nor have these efforts stopped commercial poaching. A study on commercial poaching in 1987–8 revealed that some 400 tonnes of bush meat per year reached the illegal market in Bukavu! An estimated 70 per cent of this staggering amount originated from the environs of Kahuzi-Biega National Park. If we take into account that one kilogramme of bush meat fetches some US$0.35, total revenues from this trade amount to US$140,000 per year. Added to this is the income from illegally felled trees, gold digging and poaching elephants for ivory tusks. This traffic in ivory is in the hands of both Congolese and foreigners.

Some Recommendations[2]

Most basic information on the classification of flora, the habitats of the large mammals and the ecological functions of the various parts of the park are available. It is now important to make a census of large mammals such as elephant and gorilla, and to determine their distribution. Moreover, the park hosts specific flora which need to be protected as these species have become extremely scarce in the neighbouring countries of Rwanda, Burundi and Uganda. These species include *Dicranolepsis sp.*, *Impatiens erecticornis* and *Impatiens irringii*, and mammals endemic to the region, such as *Gorilla gorilla graueri*, *Colobus polykomos* (fringed ape) and *Genetta victoriae* (giant forest genet).

As for the social aspects of conservation, the chances of achieving sustainability would be considerably enhanced if local people shared in the benefits of forest use, and if other measures were taken to ensure that their basic needs are met. A management plan that involves the local population is still lacking. Such a plan is urgently required, if only to recommend reforestation in the deforested areas. This would include secure tenure of productive farmland, guaranteed access to forest products and provision of employment. Though the new approach of IZCN and GTZ seeks to reconcile conservation with the needs of the local people, in practice customary rights over ancestral lands continue to be curtailed.

It is therefore a priority to engage in a dialogue with the local population in which the values of the forest and the problems of its conservation and use are openly discussed. In collaboration with traditional authorities, the IZCN and the GTZ should develop a new management plan that includes literacy, health care, family planning, supply of safe water and agro-forestry. This should encourage a synthesis between traditional knowledge and modern scientific insights, as it will improve the understanding between the local population and the animators who are employed by park management.

Ideally, the government should focus primarily on outreach to local communities. The GTZ could then concentrate on conservation and research activities. As a way to reach a compromise between the conservation requirements and the aspirations and needs of people, the IZCN/GTZ project should offer concrete job opportunities and other benefits for the population. This would also provide an opportunity to harmonise conservation activities with the community organisations and the traditional leaders.

Meanwhile, the government must compensate the people who were displaced following the expansion of the park and provide proper resettlement. The communities' rights over resources and their role in decision making should also be recognised in order to encourage them to protect the forest ecosystem for themselves and their children. It is crucial that tourism revenues are channelled into the provincial office of the IZCN and used to finance projects that benefit local people. In such a way, people can identify with the park and support efforts to preserve it, as the responsibility for the long-term protection of the forest ultimately rests with them.

NOTES

1 It is in the forests of this area that tens of thousands of people seek refuge from the violence between Hutus, Tutsis and other parties in the region.
2 Although the political future of the Democratic Republic of Congo is highly uncertain, the suggested actions, which have a long-term perspective, remain relevant.

BIBLIOGRAPHY

Collar, N. J. and Stuart, S. N. (1985) *Threatened Birds of Africa and Related Islands,* International Union for the Conservation of Nature (IUCN), Gland.
De Monza, J. P. *et al.* (1996) L'*Atlas pour la conservation des forets tropicales d'Afrique,* IUCN-France.

Fisher, E. (1996) *Die Vegatation des Parc National de Kahuzi-Biega*, Sud-Kivu, Zaire, F. Steiner, Stuttgart.

IZCN-GTZ [n.d.] *Parc National de Kahuzi-Biega*, Publication du Projet Zaïro-Allemand.

Lee, P. *et al.* (1988) *Threatened Primates of Africa*, International Union for the Conservation of Nature (IUCN), Gland.

Mühlenberg, M., Slowik, J. and Steinhauer-Bufkart, B. [n.d.] *Parc National de Kahuzi-Biega*.

Schaeffer, J. [n.d.] *Aspects sociaux de la protection du Parc National de Kahuzi-Biega*.

11 Forest, Water and People

In Search of a New Symbiosis for Boti Falls Forest

GREEN EARTH ORGANISATION

Ghana

Geography, Vegetation and Soil

Boti Falls is located in Yilo Krobo District, eastern Ghana, some 24 kilometres from the regional capital, Koforidua. The falls are formed at the point where the Ponpon River plunges over a 35-metre cliff. Flowing in a northwest to southeast direction, the Ponpon is hemmed in between hills ranging from 579 metres to 634 metres above sea level. It plunges over a 275-metre contour and forms 'twin falls' near the settlement of Boti Sra. The Ponpon flows over Voltaian rocks (mainly sandstone) within the Kwahu system. It flows west of the Akwapim ranges and enters the Volta Lake at a point west of Akosombo Dam, at longitude 0° 13' W and latitude 6° 11' N.

The catchment area of the Ponpon lies in the dry semi-deciduous inner-zone forest type of Ghana. The Boti Falls Forest Reserve has been classified as a moist semi-deciduous southeast type of forest (Hawthorn and Abu-Juam 1993). Some of the characteristic plants of the dry semi-deciduous inner-zone forest type which occur in great abundance are *Trilepesium madagascariense*, *Zanhoxylum leprieurii*, *Terculia tragacantha*, *Diospyros mombuttensis*, *Celtis sp.*, *Triplochiton sp.*, *Cola sp.*, *Hymenostegia sp.*, *Sphenocentrum jollyanum*, *Drypetes sp.*, *Dracena sp.*, *Antiaris toxicaria* and *Milicia sp.* Also very common are *Griffonia simplicifolia*, *Ceiba pentandra*, and *Newbouldia laevis*. The tallest trees commonly reach heights between 30 and 45 metres.

The soils of the Ponpon basin are of the Yaya-Kaple and the Yaya-Otrkpe types. At the area around the Falls, however, the Yaya-Kaple type dominates, with minor inclusions of the Pimpimso type. The soils develop over coarse-grained sandstone in the forest environment. They normally contain high proportions of ironpan and sandstone boulders, and are limited in depth. The Yaya type, for example, is susceptible to severe erosion, especially if the vegetation cover is removed. They occur on the summit and the steep slopes that adjoin the Ponpon River. Experts advocate that since these areas have little agricultural value, they should be clasified as reserve forest. The Kaple series of soils have formed on the steep lower slopes and in the narrow valley bottoms. Their plant and nutrient content is low, and susceptible to erosion. The Pimpimso type, on the other hand, has high gravel content and the soils dehydrate early in the dry season.

The soil features outlined above have clear effects on the Ponpon River. The formation of sharp slopes with high susceptibility to erosion and drought suggests that activities that lead to land degradation worsen their state. The agricultural landscape exhibits a degraded environment. In many places in the Ponpon, there is evidence of a high degree of siltation. Where the falls plunge down to forms pools, there is an abundant accumulation of sediments, dramatically illustrated by reports of swimmers getting trapped in the mud.

Climate

The nearest weather station is at Huhunya, about 6.5 kilometres from Boti Falls. For the past 15 years the rainfall pattern has shown two peak periods, in June and in October. The first rainy season is between April and June, with the second occurring in September and October. The number of rainy days ranges from 8 to 10 during these seasons. The mean annual rainfall for the area is 140 centimetres. The average relative humidity ranges between 75 and 80 per cent during the rainy seasons, and between 70 and 72 per cent for the rest of the year. The mean annual temperature is 26.7° C.

Of all the climatic elements, rainfall has been the main determinant of the Ponpon's flow, and therefore also of Boti Falls. The falls are at maximum force during the peak rainfall periods. Between December and February there is a sharp decline in rainfall; in January, the falls sometimes dry up completely.

The People and Management of the Area

The Yilo Krobo and the Manya Krobo people belong to the larger Dangme ethnic group. Boti Falls is inhabited by the Yilo Krobo, who purchased the land from the Akyem Abuakwa people at the turn of the century. During the first decade of this century, the people moved to the Boti Falls area from the plains. Currently, about 93 per cent of the population in the area is Krobo.

Traditionally, the land at Boti Falls has fallen under the jurisdiction of the Yilo Krobo Divisional Chief since its purchase from the Akyem people. The land was parcelled out in strips, each strip starting at the river, a method of land parcelling known as the *huza* system (Dickson and Benneh 1970). In 1969, however, the government of Ghana, realising the tourism potential of the falls and the fragile nature in the area, constituted 1,300 hectares of the immediate area as a reserve (Forest Department 1969).

Apart from the Divisional Chief at Huhunya, all other settlements in the Yilo Krobo area have elected leaders, known as *dademantse*. They are chosen by each community on the basis of their diligence. In recent times these leaders have been granted official status after their election by the Paramount Chief of the Krobo, who is based in Somanya, and provided with a certificate. Their duties include safeguarding the interests of the Paramount Chief and ensuring compliance with traditional sanctions for land management.

Close to the waterfall in the forest reserve there are structures built by the Eastern Regional Administration to enhance tourism. The administration of these structures has been handed over to the Yilo Krobo District Administration. Tourism amenities at the falls are owned and managed by the Yilo Krobo District Assembly in Somanya. The District Administration has agreed to pay 10,000 cedis per month to the family that owns the immediate vicinity of the falls. They also provide sheep and liquor for the ceremonies at the shrine close to the falls. The forest reserve, however, is under the management of the Forest Department, directly supervised by the District Forestry Officer.

Biodiversity in the Boti Falls Area

The dominant forest cover in the area is a teak plantation (*Tectona grandis* and *Gmelina arborea*) managed by the Forestry Department. The plantation and a small patch of secondary forest below the falls constitute the Boti Falls Reserve. The secondary forest is rather young with a broken canopy. Climbers are abundant and the undergrowth is thick. The tallest trees observed are 30–40 metres high and 2–3 metres in girth at breast height. The count of trees with a stem girth greater than 20 centimetres yielded only 11 individuals per 25 square metres.

The ecological classification of the plant species listed in the reserve indicates that the pioneers and primary species (the light demanders and the shade bearers) occur in almost equal proportions of 42 and 37 per cent respectively. The non-forest species constitute 14 per cent of the total species recorded. This indicates that if human interference is minimized the chances of forest recovery are good. About 21 per cent of the species are of immediate conservation concern. Some of these species are rare (*Crossandra sp.* and *Crinium sp.*), or under serious pressure from exploitation

(*Milicia sp.* and *Terminalia ivorensis*). The remaining 79 per cent of the species are not under threat.

The survey that was conducted shows that farming activity in the area is intensive. This has resulted in the replacement of forest with grassland, dominated by *Panicum maximum* and the weed *Eupatorium odorata*. It was also noticed that few indigenous trees, with the exceptions of *Milicia sp.*, *Newbouldia laevis* and *Ceiba pentandra*, were spared by farmers. The hill-slopes are often denuded of their forest cover and sparsely dotted with *Hildegardia sp.* Animal life in the reserve has deteriorated over the years. According to local people and the local forest guard, the bushbuck (*Tragelaphus scriptus*) is occasionally trapped by hunters. Maxwell's duiker has not been seen in the area. Squirrels and monkeys are quite common, however, and birds are abundant in an area of extensive grasslands.

Threats to Biodiversity in the Area

The inhabitants of the Boti Falls area are involved in several economic activities that have an impact on the ecosystem, including livestock rearing, wood extraction (for construction, fuelwood and charcoal) and hunting. Farming is practised at a subsistence level and involves all the inhabitants of the Ponpon valley. The main crops grown are maize, cassava, plantain, cocoyams and vegetables. Farms are usually 1.6–2.4 hectares in extent.

Shifting cultivation and population pressure

In the system of farming practised, a portion of land is farmed for maize, intercropped with cassava and vegetables. After two or three years the farmer abandons the field and moves to adjoining lands. The major factor determining the period of fallow is the amount of land available to the farmer. When there were relatively few people, farmland could be left fallow for between six and ten years, ensuring that the soil regained its fertility. Larger family sizes and land fragmentation, however, have reduced fallow periods to a typical period of three years. The soil is not replenished before being farmed again, and this leads to poor farm yields.

Use of chemicals in farming

Farmers desperate to increase their harvests (particularly those who can afford to take the advice of the agricultural extension officers) make greater use of agrochemical inputs. Chemical fertilisers and pesticides are used mostly for the cultivation of maize and vegetables. According to one of the *dademantse*, the soil is so depleted in most places that any harvest of maize is impossible without the application of such chemicals. The overuse of fertilisers and pesticides has also led to soil acidification.

Problems associated with farming methods

The slash-and-burn method of farming is widely practised in the area. The burning often gets out of control and develops into bushfires, destroying much of the flora and fauna as well as the organisms in the soil. Virtually no trees are left on the farms. Some farmers have resorted to farming in very sensitive marginal lands close to the rivers. Other farms are located in the streambeds and valleys, on steep slopes, and even (illegally) in the reserve areas. Currently, farms are located all along the banks of the Ponpon and its tributaries, some as close as three metres from the river. Most of the dry-season crops are grown in the cooler valleys of the streams, where the moisture content of the soil is high enough to sustain the crops. Unfortunately, the chemicals used in farming are harmful to the quality of surface and ground water. Farms located along the steep slopes in the area are common. As the slopes are cleared of all vegetation to make room for crop production, soil erosion accelerates. The eroded sediments and loose pebbles from these hillslopes eventually find their way to the valleys and cause siltation.

Extraction of wood

Along with farming, wood extraction is another economic activity in the Boti Falls area. Almost all valuable trees, such as *odon*, have been felled, either for commercial reasons or in the destructive course of the annual bushfires. Wood is used in the area for the construction of local huts and as building material for maize storage and livestock pens, for fuelwood and for charcoal production. Easy access to the commercial centres of Koforidua, Somanya, Tema and Accra makes the fuelwood and charcoal business very lucrative, and it is common to see heaps of fuelwood and bags of charcoal awaiting transport to the market. The removal of trees has adversely affected the groundwater regime and water retention in the soil, and has also contributed to the acceleration of bushfires and the loss of biodiversity.

Bushfires

Another human activity that has an impact on the biodiversity of the area is bushfires. Deliberate fires, for the purposes of hunting, occur during the dry season. Large tracts of vegetation are set ablaze and the escaping game are hunted down with the help of dogs and weapons. Accidental fires are started by farm fires that get out of control, or by carelessness. Although the meat of rat is taboo for the Krobo people, they hunt rats for sale to other ethnic groups. Fire is also used to gather honey, mostly for sale.

Some Solutions and Interventions

Tourism

The economic activities of people in the study area have a negative impact on the environment. But Boti Falls offers a good opportunity for eco-tourism, which could generate income for local people without affecting the biodiversity. Tourism, therefore, could offer an economic alternative and thereby reduce the pressure on the environment. To maintain the area as a tourist attraction, a continuous and sustained flow of water has to be ensured. It is also important to take measures that will improve the ecological state of the catchment area. This will mean the enforcement of legislation that makes it obligatory to set aside a 30-metre strip of land on both banks of the river. As such legislation would adversely affect local livelihoods, the authorities would have to ensure that people benefit substantially from tourism through employment with the State Tourism Services and the sale of handicrafts.

Reforestation

The original vegetation of the area was a semi-deciduous forest. To bring back a similar forest ecology, indigenous tree species must be planted. Our studies have also shown that some species are threatened or rare, which makes planting and conservation efforts all the more important. It must be stressed that the extensive slash-and-burn practice that currently prevails should be replaced by more intensive agriculture. Agroforestry and contour ploughing on the hillslopes will be necessary to offset the problems of soil erosion, the loss of soil nutrients and the removal of vegetation cover.

Woodlots

As firewood and charcoal continue to be a major source of domestic energy, woodlots will have to be established to prevent direct extraction from the forest.

Beekeeping

Another economic activity that can be combined with biodiversity conservation efforts is beekeeping. Honey extraction from the wild is widely practised by people. As the demand for honey and beeswax have increased, beekeeping has gained economic importance. Beekeeping could be introduced to people within the reserve, therefore, along with systematic training in honey and wax extraction, without the use of fire. This would minimise the prevalence of bushfires. (Community fire volunteers could also be trained and equipped to fight fires in the area.)

Game breeding

As game continues to be a source of food for the people, the introduction of game breeding would provide the necessary protein, while reducing the pressure from poaching. The meat of the grasscutter, a species of rodent, is much preferred and its breeding, which research has found to be successful, will be supported by immediate demand.

Education

Central to all the interventions is education. Although people are aware of the effects of environmental deterioration, its dynamics and repercussions are not understood fully. There is a need for extensive education to highlight the urgency for prompt action. Education for behavioural change towards sustainable living, training of local communities concerning the proposed interventions, and general awareness raising are priorities. People must be involved directly in steering the new course of action.

BIBLIOGRAPHY

Dickson, K. B. and Benneh, G. (1971) *A New Geography of Ghana*, Longman, London.

Forest Department (1969) *On the Acquisition of the Forest Reserve*, legal document, Ghana.

Hawthorne, W. D. and Abu-Juam, M. (1995) *Forest Protection in Ghana: with Particular Reference to Vegetation and Plant Species*, International Union for the Conservation of Nature (IUCN), Gland.

12 Forests for the Future

An Indigenous, Integrated Approach to Managing Temperate Watershed Resources in Oregon

DEBORAH MOORE
Environmental Defense Fund (EDF)
and
DEEPAK SEHGAL
Confederated Tribes of the Warm Springs Reservation of Oregon (CTWS)

This study of the forest management practices of the Confederated Tribes of the Warm Springs Reservation in Oregon describes the participatory planning process used to develop an integrated resources management plan for the Tribes' forests. The Tribes' reservation is located in the Pacific northwest of the United States, where controversy has been raging over the management of the area's old-growth forests, endangered species like the spotted owl and salmon, and logging jobs.

The reservation covers 260,000 hectares of central Oregon in the Deschutes River basin, about 60 per cent of which is forested with ponderosa pine and mixed conifer. Since the 1940s, timber harvesting has been a primary source of income for the Tribes. Other non-timber forest resources are culturally and economically important, however, especially huckleberries, other edible roots and medicinal plants, salmon that spawn in forest streams, wildlife, sacred sites, and recreation. Timber harvesting has run counter to other values important to the Tribes, leading to the recognition that their approach to forest management has not been environmentally, socially or economically sustainable.

The Tribal Council passed a resolution in 1986 to mandate that forests be managed in an 'integrated' manner. The resolution resulted in a multi-year process to consult with Tribal members, Tribal government staff and other interests, in order to develop a management plan that would address a broader array of resource values. The process combined scientific and technical analyses with traditional Tribal consultations and public meetings. Analyses included computer modelling and mapping of resources, while the consultations included discussions in longhouses, meetings with tribal elders, and general public meetings.

The final plan was adopted by the Tribal Council and members in 1992. It cuts the Tribes' timber harvest in half; extends protection of habitat and non-timber resources such as huckleberries, wildlife and fish; and establishes standards and best management practices to improve the conditions of lands where timber harvest will be allowed. This decision cuts the Tribes' income by about 40 per cent. The unity and consensus among Tribal members that the planning process developed will need to be continued for the implementation of the new management approach to be successful, and for alternative income-generating activities to be developed.

The Confederated Tribes of Warm Springs have collaborated with the Environmental Defense Fund since 1990. The Tribes are a federally recognised Indian tribe, comprised of the Wasco, Warm Springs, and Paiute Tribes. They signed a treaty with the United States government in 1855 in which the Tribes ceded about four million hectares of land in exchange for exclusive rights to lands and resources on their reservation. The Tribes have about 4,000 members and historically have lived throughout the Columbia River basin, following the salmon fisheries. The Environmental Defense Fund (EDF) is a US-based, national, non-profit environmental organisation started in 1967. EDF works on research and advocacy to promote environmentally, socially and economically sound policies to protect natural resources.

The Tribes invited EDF to be the only non-tribal member of the Environmental Law Team, established to assess on-reservation resource problems and recommend changes. The planning process for the forest management plan was already under way when EDF joined the team. The Tribes invited EDF's participation because they recognised that they needed advocates to promote the necessary broader reforms in the larger, non-Indian society. In addition, they felt that EDF's experience in water, land, and environmental management policies in various parts of the US would offer useful expertise to the Tribes in developing their own plans and policies. EDF's role, in part, has been to help the Tribes interact with other non-Indian interests and decision makers in the region, state, and nationally. The Tribal staff and government are responsible for interacting with the Tribal members and communities. Since 1992 EDF has collaborated with the Tribes on a new project, an outgrowth of the forest planning process, to address resource management issues off the reservation, in the larger Deschutes River basin.

The health of the Tribal economy and environment depends, in part, on the vitality of the wider region's economy and ecosystems. The challenge now is to implement and enforce the Tribes' forest management plan on the reservation, and to build the public and financial support to restore the broader environment that surrounds the Tribes' reservation. The Tribes and EDF are continuing to collaborate on a number of on-Reservation and off-Reservation activities to protect the forests, rivers and fisheries that are critical for the security and integrity of the Tribes' cultural survival.

Conflicts over Forest Resources in the Pacific Northwest of the United States

The Pacific northwest is a rugged and beautiful region of the United States covering the states of Washington, Oregon, Idaho, and northern California. It is known for big trees and salmon, which are inextricably linked via the salmon's life cycle, migrating from mountain watersheds to open oceans and back again. The Columbia River, running from Canada's British Columbia to the Pacific Ocean, is a key feature of the landscape. The region's ancient conifer forests are renowned for towering stands of trees hundreds of years old, ranging from the temperate rainforests of Washington's Olympic Peninsula to the old-growth Douglas firs of the Cascade Mountains and northern California's ancient redwoods. The forests of the Pacific northwest are the largest remaining vestiges of old-growth forest in the continental US, and are therefore of national significance.

Old-growth (or ancient) forests refer to stands of trees 1,000 years old or older. Generally, ancient forests are structurally and functionally more complex than younger forests, and support communities of plants and animals that are fully or partially dependent on old-growth habitats. In the Pacific northwest, forests are comprised primarily of Douglas fir trees; the other main species are white fir, ponderosa pine and western hemlock. These forests are home to the northern spotted owl, a federally listed endangered species, as well as bear, the bald eagle, the peregrine falcon and many other birds, mammals, plants and insects. Anadromous fish, such as salmon and steelhead, spawn in cold-water streams that flow from sources in the forests. Many of these fish species are extinct, endangered or threatened.

The original extent of forests in this region has been estimated at about 6 million hectares (Wilderness Society and Natural Wildlife 1988). Cutting of forests began in the 1800s when the first non-Indian immigrants systematically cleared and burned the forest to make way for agriculture (FEMAT 1993). In the early 1900s, commercial timber extraction began to increase, using a clearcut-and-burn approach. Logging was done by hand with saws and axes, and the supply of large trees appeared endless. So, smaller trees with low commercial value were often left standing (FEMAT 1993). After the Second World War and the invention of the gas-powered chain saw, along with other improvements in transportation, logging accelerated. European forest management methods were adopted, including clearcutting, removal of logs, snags and slash, and replanting of single species.

Today, only about 13 to 20 per cent of the total original extent of ancient forests exists (Paper Task Force 1995; Smith 1996). Most of these ancient forests are found on public lands, like national forests and Bureau of Land Management (BLM) lands; moreover, only about a third of the remnant forests (or about 5 per cent of the original area) is currently protected in parks or reserves (Paper Task Force 1995). While the total area of forest is important, so is its spatial arrangement. Much of the remaining ancient forest exists in disconnected fragments and patches, randomly shaped by geography and the pattern of clearcutting, and may not be adequate to protect forest ecosystems and their values.

Timber cutting on public lands managed by the US Forest Service and the US Bureau of Land Management has been virtually stopped by federal court orders for a variety of reasons (FEMAT 1993). Since 1972, when scientists first began to suspect that the northern spotted owl was declining because of the loss of old-growth forests, controversy over logging and forest protection has simmered. Various agencies have failed to comply with several environmental laws, including the Endangered Species Act, the National Environmental Policy Act and the National Forest Management Act (FEMAT 1993). In the late 1980s, the decline of Pacific salmon and steelhead was also linked to the loss of forest habitats. In addition to environmental constraints on logging, the Pacific northwest timber industry has faced increased competition from other timber producers, declining prices and increases in mechanisation, all of which have led to a decline in forest jobs. Many public interest groups have also begun to question the policies of subsidising logging on public lands. The 'jobs versus the environment' conflict erupted anew in 1990 and became an issue in the 1992 presidential election. President Clinton's campaign promises included resolution of the crisis over the Pacific northwest's forests.

The region's Native American communities have been particularly affected by logging and its environmental impacts. There are 25 federally recognised Indian tribes in California and 36 in Oregon and Washington that have a cultural interest or treaty rights to forests in the Pacific northwest (FEMAT 1993). These treaties and executive orders affirm certain rights, both on and off Indian reservations, to land, water, hunting and gathering, fishing, religious activities, and other activities and resources. Many of these tribes have considerable forest resources on their reservations, for which they have direct management responsibility. The Confederated Tribes of the Warm Springs Reservation in Oregon is one such tribe.

Tribal resource management activities have contributed both to problems and solutions. The Tribes' deep commitment to environmental protection has helped to forge landmark legal precedents for protecting fish, water and forest resources in the Pacific northwest. At the same time, however, many tribes have very few economic opportunities aside from the sale of timber and wood products from their reservation lands. Given the large amounts of land that tribes in the northwest control, tribal management activities can have a significant impact on the health of forests. In addition, by integrating their cultural commitment to protecting the environment with their forest management responsibilities, tribes in the northwest can develop models of sustainable management for the region as a whole, for the nation, and even for other countries.

The Warm Springs Tribes are committed to managing their resources in sustainable ways for future generations. In 1992, the Tribes asked the EDF to collaborate with them on a project to investigate and promote sustainable development and ecosystem protection strategies in the central Oregon region. The goal of the project has been to sustain both economic development and ecosystems in the Deschutes River basin, where the Tribes' reservation is located. This study describes the Tribes' process to develop 'integrated resource management plans' for resources on the reservation and the evolution towards ecosystem management for the entire Deschutes River Basin.

Warm Springs Tribes and the Central Oregon Region

In the United States, the government has generally recognised the right of Indians to remain separate and to possess and control the lands of their reservations. There are many sad chapters in the history of Indian treatment in the US, particularly during the 'termination era' of the 1950s, when policy shifted to 'assimilation' and more than 100 reservations were terminated (American Indian Resources Institute – AIRI 1988). Beginning in 1959, with the Williams court decision, the US recognised an expansive tribal sovereignty. This has evolved, through many court battles and legislation, to mean that tribes should be treated as separate, sovereign governments, similar to a state, within the US. This gives tribes broad authorities over taxation, social services and natural resources (AIRI 1988). Indian reservation lands are generally held in trust for the tribes by the US government, and the US has certain 'trust responsibilities' to manage and protect the physical, biological, social and cultural resources of Native Americans.

A Treaty of 25 June 1855, negotiated with the Tribes and Bands of Middle Oregon, the present Wasco and Warm Springs Tribes, and the Paiute Tribe which joined them in the 1870s, established the Warm Springs Indian Reservation boundaries (Moore *et al.* 1995). The reservation covers about 259,100 hectares and is located within the Deschutes River basin in central Oregon, at latitude 44° 45' N and longitude 121° 15' W. The Tribes have about 4,000 members. The Deschutes Basin covers about 10,700 square miles between the Cascade Mountains in the west, the

Ochoco Mountains in the east, volcanic lava plateaus in the south, and the Columbia River in the north (Moore *et al.* 1995).

In the 1855 treaty the Tribes ceded about 4 million hectares (about 40,000 square kilometres) to the US, while reserving exclusive use of their reservation lands. Despite ceding exclusive rights to these lands, however, the Tribes continue to have rights on, and a sovereignty that extends to, off-reservation, non-Indian areas and lands. These off-reservation rights include those to access historic fishing grounds, to burial sites and other sacred sites; to lands on which tribal members can hunt, gather food, roots, and berries, and pasture stock; and to land acquisitions in these areas. The Tribes have a constitution and by-laws for tribal government, and the tribal government now manages timber, water, salmon and other reservation resources for the benefit of its members. The Tribal Council has eight elected members and one Chief from each of the three Tribes, who is chosen for life. The Tribes own and operate the Warm Springs Forest Products Industries, as well as a recreational and tourist resort, a hydro-electric dam, and several other enterprises. Each of the three Tribes has its own longhouse, used for community meetings, ceremonies, celebrations, social services, and other functions (Moore *et al.* 1995).

The Deschutes River basin, where the Tribes' reservation is located, is a rural part of Oregon, with a population of about 140,000 people. The area has been experiencing rapid growth in recent years, due to a growing retirement community and to migration from neighbouring California. Historically, the area's economy was dominated by agriculture and logging. But in the 1990s tourism and the service sector, along with government, represent the largest employers and earnings. Lumber and wood products represent only 9 per cent of the earnings in the region, and 11 per cent of the jobs. Some of the lost jobs in the forest sector reflect a drop in timber supplies (Moore *et al.* 1995).

The Tribes own about 7 per cent of the land in the region, while the federal government manages about 49 per cent, and 42 per cent is privately owned. Federal agencies responsible for regional resource management include the US Forest Service (FS), the US Bureau of Land Management (BLM), the US Bureau of Reclamation (BOR), the US Fish and Wildlife Service (FWS), and the US Bureau of Indian Affairs (BIA). There are also several Oregon state agencies involved. The growth in the region's population and economy is largely dependent on the area's natural amenities. Tourism, recreation, and travel to the area's rivers, parks and forests are a significant portion of the economy. Consequently, the health of the region's economy depends in large part on the health of its resources and ecosystems. The Warm Springs Tribes are part of this larger region and thus partially dependent on the regional economy for their livelihoods (Moore *et al.* 1995).

Forest Resources, Forest Use and the Warm Springs Tribes

The reservation covers about 260,000 hectares, of which about 160,000 hectares are forested (about 62 per cent) (CTWS 1991). The plant associations and major timber species are: ponderosa pine (27 per cent), ponderosa pine/Douglas fir (13 per cent), grand fir (19 per cent), and Mixed conifer (12 per cent) (CTWS 1991). About 14,000 hectares are old-growth stands of 200-year-old trees. The majority of forests are between 50 and 100 years old. Receipts from the Tribes' timber sales and industries are a significant portion of their income, totalling about US$11 million annually in the late 1980s (CTWS 1992). The total volume of wood in the forested area was estimated in 1988 to be 8.5 million cubic metres. The average annual net growth per hectare is estimated to be 1.08 cubic metres, which for 158,000 hectares amounts to about 170,500 cubic metres. In 1982 the Tribal Council approved a Timber Plan with an annual allowable cut of 245,000 cubic metres (about 3 per cent of the total volume, but greater than the average annual net growth). In 1991, the Council reduced the allowable cut to 200,000 cubic metres.

In addition to timber and wood products, forests and watersheds provide many other resources to the Tribes. There are four major streams on the Reservation: the Warm Springs River, Shitike Creek, Whitewater River, and Jefferson Creek. The Deschutes and Metolius rivers form the eastern and southern borders of the reservation, respectively. In total, there are more than 2,000 miles of streams on the reservation. Water quality is generally high, and most of these rivers, except the main stem of the Deschutes, are undammed. Unfortunately, however, the reservation is located downstream of most of the river basin's non-Indian agricultural, timber, and residential developments. Upstream activities therefore greatly affect the Tribes' water resources. Reservation streams support a variety of fishing activities. Resident fish include four species of trout, suckers, and squawfish. Anadromous fish include summer steelhead, spring chinook salmon, fall chinook salmon and Pacific lamprey. The salmon species have been listed on the national endangered species list, and many programmes of the Tribes, federal agencies and state agencies are aimed at protecting and restoring salmon fisheries.

Native American cultural resources include nearly 300 sites – prehistoric, historic and cultural – which represent about 10,000 years of human use and settlement (CTWS 1991). The Tribes also consider many natural resources as 'cultural materials' because of their use in maintaining cultural integrity through traditional ceremonies, food, fibre and medicines. These plant species include: wild celery (*Lomatium nudicaule*), desert parsley (*Lomatium grayi*), biscuitroot (*Lomatium sp.*), bitterroot (*Lewisia rediviva*), blue camas (*Camassia quamash*), Indian carrot (*Perideridia gairdneri*), huckleberries and blueberries (*Vaccinium sp.*), chokecherries (*Prunus sp.*), Oregon grape (*Berberis sp.*) and black lichen (*Bryoria fremontii*). Two religious feasts of thanksgiving based on important native foods are observed every year, including the Root Feast in the spring and the Huckleberry Feast in the late summer. Feasts are also held to give thanks for salmon. Fuelwood is also considered a cultural material, since alder, larch and ponderosa pine are used in sweat lodges and other traditional activities. With population growth, increased interest in cultural heritage, increased reliance on subsistence plant and animal food sources, and shrinking availability of these resources in adjacent lands, it is estimated that demand for these resources will increase. Also, a few Tribal members have begun to collect non-timber forest resources for commercial sale, including edible mushrooms, medicinal herbs, pine cones and other decorative floral materials.

Rangelands, both timbered and non-timbered, provide forage for approximately 1,600 cattle, 1,800 horses and 3,500 deer and elk (CTWS 1991). About 140 Tribal members and their families benefit from livestock grazing. The reservation also provides habitats for more than 120 different species of reptiles, amphibians, birds and mammals. Big game animals, mule deer and blacktailed deer, Rocky Mountain elk and black bear are hunted by Tribal members. Some trapping of beaver, muskrat, mink, otters and other furbearers occurs on the reservation. Interest in wildlife, particularly big game, has been increasing among Tribal members, but hunting limits have not yet been set (CTWS 1991).

As well as using the forest for their own recreation, the Tribes also own the Kah-Nee-Ta Resort, which provides similar opportunities to visitors. Recreation includes hiking, camping, fishing, swimming, rafting, hunting, boating, skiing, off-road vehicle use, nature studies and cultural food gathering. The Tribes generate income via the Resort and via fishing permit fees. Other recreational activities could generate revenue, and these opportunities are being investigated.

As described here, the forest resources of the Tribes are extensive. Timber and wood products have been a significant source of income for the Tribes, but the forest provides many other valued resources. Of particular importance are the wildlife, food and medicinal herbs. These resources are not simply 'products', but part of the Tribes' religion and culture. For these people, who have lost so much of their culture and way of life since European settlement, sustaining salmon runs, native plants like huckleberries, and bear, elk and deer is a necessary dimension of their cultural survival.

Tribal Forest Management Approaches: Past to Present

In pre-European periods, the three Tribes used resources from numerous areas spread across the northwest. They would move seasonally, following weather and salmon runs. Plants, roots, and berries would be harvested in the summer in the mountain forests. Fish would be harvested at different sites throughout the Columbia River basin during spring, summer and autumn. Shelter was made primarily from grasses and skins, with some use of wood for poles.

Commercial logging was introduced on the reservation in the 1930s, but did not become a significant enterprise until the 1940s and 1950s. The main objectives in this period were to provide income to the Tribes, develop roads in the reservation, and remove old, diseased, slow-growing trees from the stands (CTWS 1991). The predominant harvest method was selective cutting of mature trees, focusing on ponderosa pine. Selective cutting of pine, combined with exclusion of wildfire, led to a change in species composition with mixed conifer stands now predominant.

During the 1960s and 1970s selective cutting continued. In mid-elevation and higher-elevation forests, however, clearcuts, shelterwoods, commercial thins, and partial cuts were the predominant harvest methods. Shelterwood is a forestry system used to harvest mature trees at rotation age in a series of preparatory, seed and removal cuts designed to naturally regenerate a new even-aged crop of trees under the shelter of the old crop. As demand for wood products grew in the 1970s, the understorey of Douglas fir and white fir was also increasingly harvested. Roads were forged through forested areas, as well.

Starting in 1982, timber harvesting began to be managed under the Forest Management Plan, which was approved by the Tribal Council and the Bureau of Indian Affairs. This change was, in part, due to the Tribes (instead of the BIA) exercising greater control over their own resources. For the first time, an 'annual allowable cut' was established with the goal of managing harvests for long-term timber production. The allowable cut was set at 245,000 cubic metres. This plan has been revised three times since to reflect the Tribal Council's policy momentum for the protection of resources. In 1984, and twice in 1987, the allowable cut was reduced and several thousand hectares were removed from the commercial timber base because they were environmentally sensitive. As part of this change in approach, the Tribal Council adopted Resolution 7410 in 1986, which called for the use of an 'integrated planning approach' in the development of all future resource management plans. The next several years were spent developing the Integrated Resource Management Plan for the Forested Area.

The Integrated Resource Management Plan

The Warm Springs Tribes are at the centre of many seemingly contradictory forces, which they have to resolve in order to make decisions and move forward. On one hand, they have a culture, tribal elders, Chiefs and traditions that lead them in the direction of traditional, subsistence uses of forest and watershed materials. On the other hand, they live and operate within the broader context of a 'modern', technically driven, industrially-based economy. They must reconcile their beliefs and traditions with the economic realities and scientific uncertainties of natural resource management. The process they used for developing the Integrated Resources Management Plan (IRMP) has achieved a delicate balance between these forces and has won consensus in the Tribes' communities.

Purpose and Philosophy of the Planning Process

The purpose of developing the management plan was to produce guidelines for the use and protection of all forest resources and to establish a basis for making decisions. The ultimate goal was to provide for the economic and cultural security and health of the Tribes and Tribal members of both present and future generations. To ensure that the range of resource management options that were available then remained available to future generations, the Tribes' goals were to promote the long-term productivity, health and stability of the total forest ecosystem. They wanted to protect resource values, while providing key products such as timber, fish, forage, wildlife and water.

To develop the plan, the Tribal Council appointed an Interdisciplinary Team (ID Team) in 1988, composed of staff from Tribal resource departments and from the BIA. A ten-year planning period was chosen (1992–2001), with longer-term effects incorporated into the evaluation process. The ID Team worked with the Tribal Council, Committees, departments' staff, members, the BIA, and other outside experts to identify issues and concerns. A resource management questionnaire was sent out to all Tribal members to survey opinions about important issues. Public meetings for Tribal members to talk were held at longhouses, at the Kah-Nee-Ta Resort, and at other locations. Meetings were held with Tribal government staff and with Tribal committees. Input from the BIA was also solicited via questionnaires and meetings.

After all this participation, 13 major issues and concerns had been identified that the planning process needed to address (the issues are described in more detail below). An inventory of the Tribes' resources was also done, to provide baseline information about the availability and conditions of different resources (summarised above). Based on the issues and resources identified, eight 'management zones' were developed, which were used to help describe the multiple capabilities of different forested areas and to design management options. These eight zones were: timber, wildlife, riparian areas, rural housing, conditional use areas, culture, recreation, and biological diversity islands. Standards for land stewardship that apply to all management zones were developed, as well as 'best management practices' (BMPs) that apply to specific management zones. Lastly, four alternatives were developed to provide the Tribes with a basis for determining direction in the 1992–2001 period.

The four alternatives provide a range of outputs and environmental effects:

1 *No action* – Continuing management according to the 1982 plan, which emphasised timber production.
2 *Timber* – Providing a higher level of timber revenue to the Tribes, without neglecting full protection of cultural resources and adequate environmental protection,
3 *Balanced* – Providing a high level of benefits for Tribal members from all resources, balancing monetary returns with other values.
4 *Amenity* – Providing high levels of quality wildlife habitat, biological diversity and environmental quality.

The Tribal members and Tribal Council needed to choose among these alternatives to establish a basis for directing the various agencies and industries in their day-to-day operations. An environmental assessment (EA) of all the alternatives was prepared to assist in decision making. A number of computer-based tools were used to organise, display, and analyse all the information collected. Programmes were used to process the forest inventory data, to estimate the future yields of timber stands, and to estimate the allowable cut for the different management options developed. Satellite imagery was used to identify and evaluate wildlife habitats. A computerised Geographical Information System (GIS) was used to organise and display map information, and to assist in evaluating the cumulative effects of different activities occurring within watersheds.

Tables 12.1 and 12.2 present summaries of the effects of the four alternatives. Table 12.1 shows how lands would be allocated under the different alternatives. For example, the amount of land allocated to timber harvest would be decreased from 127,000 hectares in the 'No action' alternative to 58,000 hectares in the 'Amenity' alternative.

Table 12.2 shows the output or effect of the different alternatives. For example, the annual allowable cut decreases steadily from the 'No action' alternative to the 'Amenity' alternative. The 'Balanced' alternative does not produce the highest environmental protection or the highest timber revenues; it balances the trade-offs.

Ultimately, after another round of public meetings, longhouse

Table 12.1 Management zone allocations by alternative (allocation in hectares)

Management zone	No action	Timber	Balanced	Amenity
Conditional use	26,3221	24,6682	24,6682	24,6682
Extensively managed forest[1]	0	2,943	2,943	2,943
Extensively managed forest[2]	0	0	3,022	3,022
Timber	126,796	119,347	75,609	58,213
Biodiversity islands	0	0	8,313	14,268
Wildlife	0	0	28,975	40,415
Water resources Riparian zones A & B[3]	1,707	8,590	11,101	11,101
Recreation	773	0	918	918
Visuals	180	180	180	180
Trails[4]				
Housing[5]	0	0	0	0

1 Includes Riparian Zone A.
2 Does not include Riparian Zone A.
3 Riparian zones A & B within the forest.
4 Estimates only.
5 For illustration – do not add in for total reservation acres.

Table 12.2 Outputs and effects by issue and alternative, Warm Springs Reservation

Output/Effect	Issue	Units of measure	No action	Timber	Balanced	Amenity
Fuelwood	4	Cords	2,900	2,000	1,900	1,700
Post and poles	4	M pieces	5	5	4	3
Christmas trees	4	M Trees	50	50	50	50
Cultural plants	7	Hectares	17,111 Huckleberry and 1,630 other	Same	Same	Same
Biological diversity:						
Unmanaged growth	8	Hectares	5,704	5,704	5,704	5,704
Managed old growth	8, 1	Hectares	0	0	1059	1059
Species diversity	8, 1	Trend	Low	Low	Medium	High
Forage production	9	MAUM/ year	50.0	62.5	60.0	58.5
Annual allowable cut	10, 13, 6, 3	Cubic metres	200,000	134,000	127,000	120,000
Overall water quality	11	Trend	Declines	Maintained	Improves	Improves
Deer population	12, 5, 8	Deer	2,000	3,000	6,000	7,000+
Elk population	12, 5, 8	Elk	200	300	700	1,000
Anadromous fish	12, 11, 5, 8	Trend	Declines	Stable	Increases	Increases
Budget (estimate)	3	Million US$	3.0	3.0	3.3	3.6
Staff	3	Number	75	75	82	90

discussions and salmon feasts, the Tribal members and Tribal Council adopted the 'Balanced' option. The Tribal Council passed Ordinance No. 74 on 29 April 1992, adopting the IRMP and 'Balanced' option and phasing in the reduction in timber harvest over a five-year period from 1992 to 1996. The elements of the plan are outlined below.

Elements of the Integrated Resources Management Plan for the Forested Area

The general elements of the selected alternative combine and optimise goals for environmental protection, cultural traditions and subsistence resource uses, timber harvest and revenues, and recreation and housing needs. The forested areas of the reservation were allocated among different management zones and activities. For example, in areas allocated to timber production, a full harvest would be promoted and intensive management for timber would occur. In areas designated for wildlife, a partial harvest would be allowed, and in biological diversity islands no harvest would be allowed.

For the 'Balanced' alternative selected, the planners developed goals, objectives and desired future conditions for 21 separate resources from soils to timber health and fisheries. Based on these objectives, the IRMP establishes standards and BMPs that apply to the whole forested area, as well as special standards and BMPs for specific management zones (CTWS 1992).

For example, CA13S is a forest-wide standard for cultural resources to initiate a programme to encourage the return of cultural plants in areas where they are diminishing. A forest-wide BMP was established, CA11, to work with the Tribal Forestry Department to develop an ongoing pro-gramme to promote timber harvest and site preparation activities that are compatible with huckleberry regrowth and berry production, such as planting huckleberry seedlings in harvest areas. Then, for specific management zones designated for cultural materials, additional standards and BMPs were set, such as ZCAS1, requiring a 30-metre buffer around designated root-collecting areas, and ZCA9, requiring that timber harvest be avoided in designated areas of black moss. Hundreds of standards and BMPs were set for the forest as a whole and for the eight management zones, covering activities such as logging practices, erosion control, road construction, treatment of tree slash, water resources and fuelwood cutting.

For timber and logging activities in particular, many new rules were established. First and most significantly, the area allocated to intensive timber harvest was reduced by nearly 50 per cent. The annual allowable cut was also reduced by about 40 per cent, from 200,000 cubic metres to

127,000 cubic metres, which will be phased in over a five-year period. Second, timber harvests and the type of logging practices allowed are controlled in the areas allocated to other resources, such as biological diversity islands or wildlife. Third, specific standards and BMPs were set to control logging practices. These include limiting the size of clearcuts to 16 hectares, requiring logging plans, protecting specific trees and stands that are valuable for habitat, limiting skidding distances, increasing the rotation length (up to 250 years in some areas), and requiring buffer strips in riparian areas along streams, among others.

The IRMP lays out the plan at a broad level for the larger forested area. The Tribes also developed a second planning process for individual activities within a specific project area, such as timber sales. There is now a planning, project development and project approval process, to ensure that specific projects fit with the overall IRMP. Lastly, the Tribes developed implementation, enforcement, monitoring and evaluation mechanisms. The Tribal Council and the Tribal Chief Executive Officer have primary responsibility for Tribal management actions. They can delegate authority to the general managers of different departments (such as Natural Resources). They can also make recommendations to the BIA and the Secretary of the Interior Department (a member of the President's Cabinet), who have trust responsibilities to the Tribes. The Resource Management ID Team is now a well-established body, which meets annually, to evaluate implementation and report on accomplishments to achieve the goals of the IRMP.

Economic, Social, and Environmental Consequences of Implementing the Plan

The IRMP process and plan is a classic example of 'planning', in the sense of establishing goals, objectives, and management programmes to achieve the goals. Yet, the art of what the Warm Springs Tribes have done is the integration of multiple values, from both their traditional and modern cultures, in the context of a planning process. Difficult choices were faced at every step of the process.

Table 12.3 shows the economic 'trade-off' analysis of the four alternatives evaluated during the planning process (CTWS 1991). The 'No Action', or business-as-usual, alternative would have produced $10.8 million in timber revenue and 'costs' of almost $500,000 in forgone timber revenue for protecting non-timber resources. For the chosen 'Balanced' alternative, the Tribes' projected timber revenue would be $6.9 million (a reduction of almost 40 per cent from the recent average of $11 million), while the 'cost' of forgone timber revenue to protect non-timber resources was $1.2 million (more than twice as much as in the 'No Action' alternative). Since profits from Tribal enterprises are generally distributed to each Tribal member, the reduced timber revenues represent a significant drop in individual incomes, as well as lower revenue for Tribal government and social, education, and resource management services. The number of staff employed in the 'Balanced' option will actually increase (from 75 to 82) because more 'resource managers' will be employed, not only loggers and wood processors.

Table 12.3 IRMP timber harvest trade-offs analysis, Warm Springs Reservation

Management zone	'No action' alternative	Timber alternative	Balanced alternative	Amenity alternative
Projected timber revenue[1]	$10,809,240	$7,253,064	$6,894,888	$6,498,336
Projected timber revenue trade-offs[2]				
Wildlife/ Reg	$0	$0	$39,408	$101,897
Wildlife/ Term	$0	$0	$12,436	$24,166
Biodiversity islands	$0	$0	$308,095	$528,845
Visuals	$22,272	$0	$28,301	$ 28,301
Riparian				
Zone A	$5,015	$264,610	$264,610	$264,610
Zone B	$0	$0	$77,368	$77,368
Conditional use areas	$462,062	$462,062	$462,062	$462,062
Extensive management	$0	$0	$3,166	$3,166
Forage	$0	$0	$7,409	$15,356
Cultural	$3,084	$3,084	$3,084	$3,084
Total timber revenue trade-offs	$492,433	$729,756	$1,205,965	$1,508,855

Notes: 1 Timber harvest revenue was calculated using a stumpage price that represents the projected price for the mid-point of the planning period. Prices were developed by projecting the 1989 appraised prices forward with an annual 3 per cent real rate of increase. The species breakdown of the projected timber harvest is Ponderosa pine = 23 per cent, Douglas fir = 39 per cent and White fir = 38 per cent.
2 Timber harvest revenue trade-offs represent the opportunity cost (in terms of timber revenue forgone) of constraining timber harvest to satisfy the Tribal objectives pertaining to non-timber resources.

Overall, the environmental effects of the 'Balanced' alternative are positive. While the commercial timber harvest, using modern, industrial silvicultural methods, will continue on reservation forests, this alternative represents a significant change in approach. The Tribes have cut in half the amount of forest to be used for timber. This alone will have positive environmental benefits, since the forest that was previously open to logging will now be managed for other resources. Also, the volume of trees cut (not only the area from which they are harvested) was reduced by about 40 per cent, which reduces the need for clearcuts and other intensive practices. In addition, the standards and BMPs established as part of the IRMP will ensure that better harvest and management methods are used for all forest lands. For example, limiting the size of clearcuts, requiring buffer strips, and limiting how logs can be skidded and transported will all help to protect soil stability, stream flows, water quality, and wildlife habitat. Deer and elk populations are expected to increase at least threefold, and salmon and steelhead fisheries are expected to increase. The monitoring and evaluation of the IRMP's implementation will provide information on whether these positive environmental effects are fully realised.

The social impacts are also generally positive. Tribal members support the plan and are pleased with the process used to develop it. There is general consensus about the need to sustain the forest's resources in the long term. Even though the economic consequences of reducing timber operations are painful, the support in the community helps to fuel the development of alternative commercial enterprises. Tribal members, particularly the elders and women, appreciate the attention given to sustaining cultural materials and are excited by programmes to increase huckleberry production. Everyone is enthusiastic – at a very deep, spiritual level – about the commitment to restoring salmon populations, since salmon are the basis of the Tribes' traditions, religion and culture.

The four main enterprises being explored to replace timber revenues are: (1) greater tourism and recreation; (2) expansion of milled wood products, rather than simply raw logs; (3) other manufacturing, such as an expansion of existing clothing manufacturing; and (4) gambling. The Kah-Nee-Ta resort has been expanded and some other recreation areas have been opened to non-Indians. The Warm Springs Forest Products Industries already produces some milled and processed wood products and now plans to expand, as well as re-tool, to be able to process smaller logs. Similarly, the Warm Springs Clothing Co. has plans to expand.

Gambling on the reservation is more controversial. Gambling is illegal in many states, but Indian tribes' sovereign status has given them the opportunity to develop gambling enterprises on reservation lands. While gambling can offer major profits, the Warm Springs Tribes have resisted this opportunity for decades because they do not view it as a truly 'productive' enterprise. They are now experimenting with a limited gambling operation, to be located on a main road to Portland, Oregon's biggest city, but away from the reservation's main towns.

Threats to Sustainability

The IRMP provides the Tribes with a basis for making decisions. In and of itself, it is not enough to achieve sustainability of the reservation's forests.

Factors internal and external to the Tribes can affect the chances of successfully protecting the forest. Within the Tribes, there are differences in perspective on how to proceed. For example, the decision to pursue gambling as an alternative income generation activity has met with a mixed reception. Different generations have different priorities, with the oldest generation – and significantly, some of the generation now in their twenties – more committed to environmental and cultural protection. Some of the middle-aged Tribal members are more concerned with the economic well-being of the Tribes, which is also a valid and important concern. To the extent that unity and consensus has been a strength in the planning process, the possibility of disunity could affect the ability of the Tribes to implement the integrated plan fully. If monitoring and enforcement are not pursued, then the plan will not be successful in changing on-the-ground logging practices. Also, if income is reduced too drastically and alternative sources are not successful, there will be greater pressure to increase revenues from logging.

External to the Tribes, there are numerous local, regional, state, national and global trends that can affect the implementation of the IRMP. For example, national efforts to enforce the Endangered Species Act, a law which the current Republican congress is trying to weaken, could stop more logging on national forests and other public lands. This could benefit the Tribes by increasing stumpage prices, thereby making the timber they produce more valuable. Endangered species protection could also force the Tribes to curtail their own activities more than they have planned, however, since spotted owl and other endangered species do occur in Reservation forests.

Other factors include the Tribes' relationships with other state and federal agencies, such as the US Forest Services, the BIA, and others. Federal agencies have a 'trust responsibility' to work with tribes to protect their resources. These agencies also have other mandates, however, which sometimes conflict with these trust responsibilities. The trend has been for the Tribes' needs to be lower priorities than the other missions of the agencies. The Tribes' efforts to restore salmon fisheries – a goal of the IRMP – may be hindered if the other agencies fail to support their efforts. The US Forest Service has timber operations in areas adjacent to or near the reservation which affect salmon habitats. Yet, the Forest Service has failed to comply with many requirements to protect salmon habitat, in favour of logging. This is an example of how the actions of other agencies may inhibit the Tribes' success.

The global market for wood and wood products can also affect the Tribes' efforts, depending on future wood supplies, demands, prices and labelling requirements. If global supplies are plentiful and prices are driven down, the economic consequences for the Tribes could be worse than predicted, which may, in turn, pressure them to log more. If the demand grows for 'certified wood' – wood that was harvested sustainably – the Tribes could try to receive certification of the IRMP and their harvest methods. This could have positive or negative economic consequences, depending on how much they might have to modify the IRMP and current harvest practices.

Additional Actions

The IRMP has been the first step in addressing the Tribes' environmental and resource concerns. This process has addressed the forested area on the reservation, one of the primary resources the Tribes depend on. Plans are already under way to address other resources, both on and off the reservation. On the reservation, a planning process has begun to address toxic and hazardous materials. In addition, the Tribes are negotiating with state and federal governments regarding the fulfilment of the Tribes' water rights, which have not yet been quantified. When a settlement is reached it needs to be ratified by legislation from both the state legislature and the national Congress. Negotiations have been going on for more than a decade and are very close to being finalised.

The Tribes have also initiated several efforts to address off-reservation resource issues within the Tribes' ceded territories throughout the Columbia River Basin. The Warm Springs Tribes are working with the four other Native American communities that have treaty rights to salmon fisheries on salmon restoration via the Columbia River Inter-Tribal Fish Commission (CRITFC). CRITFC has worked to stop or modify logging operations on national forests that affect salmon fisheries, and has succeeded in having some of the hydroelectric dam operations changed in favour of fish passage.

Within the Deschutes River basin, the Tribes invited the EDF to collaborate on developing a programme for environmental restoration. This effort started in 1992 and has involved research into specific water-resource and land-use problems. The project recently culminated in the introduction of legislation in the US Congress by Oregon's senior senator, Mark Hatfield. The bill – the Deschutes Basin Environmental Restoration Act (Senate Bill 1662) – has provisions to change logging, grazing, irrigation, hydroelectric and urban development activities in keeping with specific environmental restoration goals. In addition, the bill proposes methods for public and private cost-sharing of the expenses of environmental programmes, such as user fees on recreational activities, fees on pollution discharges and new urban developments, and fees on logging, grazing, irrigation and hydroelectric operations. The bill has passed the Senate and is awaiting a vote in the House of Representatives.

Finally, the Tribes are involved in a new project with EDF to investigate the potential for the forests' other ecological benefits to produce economic value (EDF 1996). For example, the forest has value for sequestering carbon dioxide, the atmospheric gas causing global warming. Yet this value is often not realised. Other economic sectors, such as the electricity industry, are being forced to reduce carbon dioxide emissions and/or increase the absorption of carbon dioxide. These industries could pay forest owners for the ecological benefits of maintaining intact forests as one step towards fulfilling their responsibilities for reducing carbon dioxide. In this new project, the Tribes and EDF intend to investigate the science and economics of 'marketing' these ecological 'products', which could provide another alternative source of revenue.

Several outside efforts are also needed for the Tribes' forests, other resources, and cultural and economic security to be sustained. More support from other local, state, regional and national governmental entities is needed. Indeed, these agencies should be following the Tribes' example and creating better integrated plans for managing other public forest resources. Such changes would benefit both the Tribes and non-Indian communities. Furthermore, national agencies have trust responsibilities to the Tribes that are not being fulfilled, such as protection of huckleberries and other cultural materials on US Forest Service lands. The US government needs to fulfil its treaty obligations and collaborate with the Tribes on implementing management plans to protect their treaty resources. Also, much of the Pacific northwest's remaining forests are in private lands: the Tribes' experiment in balancing profits with other resource values is an example to many private timber companies and landowners, as well as to public agencies.

Conclusions and Recommendations

During the last decade, the Confederated Tribes of Warm Springs have engaged in an in-depth process to develop and implement an integrated plan to manage the forests on their reservation. This process culminated in 1992 with the adoption by the Tribal Council and the Tribal members of the IRMP for the Forested Area. The most dramatic change, as we have seen, was to reduce by 50 per cent the area of forests open for logging, thus reducing the Tribes' revenues and income by about 40 per cent.

This dramatic decision was made possible by the open, engaging, and sensitive process used to involve Tribal members in identifying the key issues of concern and in selecting the appropriate management alternatives. Detailed resource inventories, data analysis and evaluations were made of the four possible alternatives. The process showed the planners' and communities' commitments to incorporating traditional values into a scientifically sophisticated analysis. The final plan for managing the forests includes enhancing many resources of crucial cultural importance over timber revenues, including the expansion of huckleberry production, the protection of sacred sites and the restoration of salmon fisheries.

Historically the Tribes have worked by consensus, and this planning process continued that form of decision making. This collective support is critical to the long-term success of protecting their forests, particularly in the face of renewed economic hardships brought about by the decision to reduce timber harvests. If the support of Tribal members diminishes, implementation of the IRMP will most likely fall short of expectations.

Several strategies could advance the efforts of the Tribes to sustain their forests:

- Continuing to develop alternative income-generation sources such as tourism and recreation, processed and finished wood products, other manufacturing and, possibly, gambling. These efforts will need the support and investments of neighbouring communities, state and federal agencies, and the private sector.
- Continuing the processes of involving Tribal members and communities in the ongoing decisions about managing the forests and other natural resources on the reservation through longhouse meetings, public hearings and other forms of participation.
- Engaging other interest groups, particularly in the Deschutes River basin, to develop support for environmental restoration programmes and to assist in raising funds for these, such as the fees proposed in the Deschutes Basin Environmental Restoration Act legislation.
- Rallying political support to pass the Deschutes Basin Environmental Restoration Act and to force the state and federal agencies to fulfil their trust responsibilities to the Tribes and to comply with environmental laws. Such support can be developed by building partnerships with other Indian groups, other environmental organisations and the general public, especially in Oregon.
- Sharing the process and results of the IRMP for the Forested Area with other Indian communities that have forests, with private forest landowners and companies, and with state and federal forest agencies. Since the health of the Tribes' wildlife and ecosystems depends, in part, on the vitality of neighbouring ecosystems, the Tribes will depend on others – both public and private entities – to protect their forests, too.

The Tribe's IRMP can serve as a model of how to change the approach to forest management. Earlier forest practices on the Warm Springs Tribes' Reservation were not aimed at protecting the forest and its ecosystems in the long term. Through an open and participatory process, they were able to make difficult and critical decisions to shift towards a more sustainable forest management strategy that protected their culture as well as their forests. Their example may encourage others to reduce logging and focus on the non-timber resources and values that forests produce.

BIBLIOGRAPHY

The American Indian Resources Institute (AIRI) (1988) *Tribal Water Management Handbook*, Oakland, California.

Checchio, E. and Colby, B. G. (1993) *Indian Water Rights: Negotiating the Future*, Water Resources Research Center, The University of Arizona, Tucson.

Confederated Tribes of the Warm Springs Reservation of Oregon (CTWS) (1991) *Environmental Assessment of the Integrated Resources Management Plan for the Forested Area*, Confederated Tribes, Warm Springs, Oregon.

Confederated Tribes of the Warm Springs Reservation of Oregon (CTWS) (1992) *Integrated Resources Management Plan for the Forested Area, January 1, 1992–December 31, 2001*, Confederated Tribes, Warm Springs, Oregon.

Environmental Defense Fund (1996) *Pacific Northwest Forest Program: Summary and Description*, Environmental Defense Fund, New York.

Forest Ecosystem Management Assessment Team (FEMAT) (1993) *Forest Ecosystem Management: an Ecological, Economic and Social Assessment*, US Forest Service and the Bureau of Land Management.

Moore, D., Willey, Z., Diamant, A., Calica, C. R. and Sehgal, D. R. (1995) *Restoring Oregon's Deschutes River,* Environmental Defense Fund and The Confederated Tribes of the Warm Springs Reservation of Oregon, New York.

The Paper Task Force (1995) *Environmental Issues Associated with Forest Management*, Environmental Defense Fund, New York.

Senate Bill 1662, *The Deschutes Basin Environmental Restoration Act*, sponsored by Senator Mark Hatfield, Republican from Oregon.

Smith, G. R. (1996) *Principles of Ecologically Based Forestry in the Pacific Northwest*, Environmental Defense Fund, Bend, Oregon.

The Wilderness Society and National Wildlife Federation (1988) *Pacific Northwest and Wood Products: An Industry in Transition*, The Wilderness Society, Washington, DC.

13 Working with the Woods

Restoring Forests and Community in New Brunswick

MATTHEW BETTS and DAVID COON

Conservation Council of New Brunswick (CCNB), Canada

In Canada, responsibility for forests falls within the constitutional jurisdiction of the provinces. New Brunswick is Canada's most forest-dependent province with 85 per cent of its landscape classified as forest land. It is also the province with the longest history of intensive commercial logging, dating back to 1805. The province encompasses the traditional territories of the Maliseet people, and the western portion of the Micmac territory. It is currently subject to land claims as title was neither transferred by treaty nor seized by conquest. Half of New Brunswick's forest is Crown land licensed to eight companies, six of which are multinationals that own pulp mills in the province. Twenty per cent of the forest is owned outright by pulp companies, primarily J. D. Irving Ltd, Noranda and Georgia Pacific. The remaining third of the forest is owned by 41,000 small woodlot owners.

It is difficult to overestimate the importance of the forest to the people of New Brunswick. It has shaped the province's culture, politics, social relations and economy. For many in rural and small-town New Brunswick, winter is the time to cut firewood for the following year. Collecting sap from maple trees to make maple syrup heralds the first days of spring. Later, when the last snows disappear from the woods, the fiddlehead (fern shoot) harvest starts along New Brunswick's river and stream banks. As the warmer weather arrives, the salmon return to the province's rivers to spawn, and fishing begins. Before the snows return, many head to the woods to hunt for deer – a source of inexpensive meat for the freezer.

By the turn of the nineteenth century, New Brunswick had become extremely dependent on the timber trade; a situation that persists to this day (Lower 1973). The forest industry employs approximately 12,000 people directly in the pulp, paper and sawmill sectors. An additional 7,500 earn their living indirectly from the forest industry (Forest Products Commission 1994, p.1). Many of the more than 41,000 private woodlot owners in the province depend on the forest for income and fuelwood. It is estimated that one out of every eight New Brunswickers is employed directly or indirectly in the forest industry. The forests are thus the economic mainstay of most of our communities.

Unfortunately, New Brunswick's forests have not been managed sustainably and often the interests of the province's citizens have been ignored. Since the health and well-being of so many New Brunswick communities are intimately tied to the health of our forests, as they become degraded, the human settlements and activities which depend upon them inevitably decline. The symptoms of an emerging ecological, economic and social crisis are clearly evident in the New Brunswick forest sector. Over the past 30 years, employment in the logging and wood-processing sectors has plummeted (Statistics Canada 1995). Natural forests are being liquidated to supply the massive fibre needs of the pulp and paper industry. Forests are young and fragmented, forest stands have become simplified, logging roads are extensive, and an increasing number of native flora and fauna are in decline. Meanwhile, only one per cent of the province's forests are protected from exploitation, and prospects for the creation of more representative protected areas, crucial to the maintenance of biodiversity, are bleak (*New Brunswick Reader* 1996).

The dominant position of the pulp and paper industry in the provincial economy shapes both provincial politics and the politics and social relations of forest-based communities. Communities concerned about their future, forest workers concerned with declining employment, and citizens concerned about the degradation of our forests have few opportunities to shape forest policy (CLURE 1993). To move beyond this imminent crisis, alternatives to the *status quo* in forest policy need to be found and tested. The Conservation Council is a citizen-based environmental organisation that has been campaigning for a transition from centrally planned forest use to community forestry. Several communities in New Brunswick are already experimenting with community forestry. If the concept is to benefit the province as a whole, however, it must be tested on a broader level. This will require more flexible forest legislation, considerable political will, and a knowledgeable and enfranchised public.

The Forests of New Brunswick

New Brunswick's forests are part of the unique 'Acadian' mixed forest zone: spruce-fir and northern hardwoods (the exception is the boreal forest covering the ancient Appalachian Mountains which range through the north-central part of the province). This is a zone of vegetational transition between the coniferous boreal forests of the north and the broadleaf deciduous forests of the eastern United States (Wynn 1981). Today, over 86 per cent of New Brunswick is considered to be forested, making it the most forested province in Canada (Forest Products Commission 1994), but this statistic hides the tremendous changes that have occurred in New Brunswick's Acadian forest since timber exploitation began in the early 1800s. New Brunswick's woods have been transformed by various forms of human disturbance, ranging from land clearing for farms and settlement, to relentless 'mining' of certain tree species.

As late as 1800, the forest's characteristics had not been altered much by European settlers (Wynn 1981). New Brunswick's forests were tremendously diverse, with a total of 32 different species of trees which contributed to a wide variety of forest types. Hardwood ridges, mixed forests, red spruce (*Picea rubens*), a major source of spruce gum, white pine (*Pinus strobus*), paper birch (*Betula papyrifera*), yellow birch (*Betula alleghaniensis*) and alder stands (*Alnus rugosa*), and 'barrens' might have been encountered within a kilometre of each other (Wynn 1981). The cougar (*Felis concolor*) and grey wolf (*Canis lupus*) were the top carnivores and caribou (*Rangifer tarandus*) and moose (*Alces alces*) were the most common ungulates.

Another distinguishing characteristic of early New Brunswick forests was the size and age of the trees. Historical documents suggest that common shade-tolerant species such as beech (*Fagus grandifolia*), sugar maple (*Acer saccharum*), red spruce and hemlock might live to be 200–400 years old (Perley 1847). Graeme Wynn (1981), in his historical work on the province, lists six broad forest zones that are remarkably different from the forest cover of today. These included a deciduous zone, dominated by white ash (*Fraxinus americana*), butternut (*Juglans cinerea*), ironwood (*Ostrya virginiana*) and basswood (*Tilia americana*), all of which were relatively uncommon elsewhere in the province; other zones grouped sugar maple, red spruce, hemlock (*Tsuga canadensis*), and pine; sugar maple, ash; red spruce, hemlock, pine; sugar maple, yellow birch and fir; and fir, pine, and birch. Other forest communities that were regionally significant included those dominated by such species as eastern white cedar.

The Historical Development of Forestry in New Brunswick

A knowledge of the history of forestry in New Brunswick is crucial to understand the emerging crisis in New Brunswick's forests and society. Over two hundred years of forest exploitation have left their mark. An industry that is controlled by a few large economic interests, a government that is by and large 'captured' by these interests, degraded forests, and disenfranchised, powerless communities are not new developments. Nor are they unrelated phenomena. They have their roots in circumstances and policy which originated in the early 1800s.

At the beginning of the nineteenth century, New Brunswick was still essentially an agricultural society. Forest exploitation was limited by the rudimentary technology available (oxen and axe) and labour tended to be irregular and unspecialised. Individuals usually held many occupations. This situation changed radically when Napoleon blockaded Baltic ports in 1805, cutting Britain's traditional supply of timber and triggering a timber supply shortage of crisis proportions. Demand for British North American wood skyrocketed (Lower 1973). Between 1805 and 1812 New Brunswick's exports of ton timber increased almost twenty-fold (Wynn 1981, p. 33).

Shrinking markets in the early 1900s contributed to the collapse of the lumber industry and hence the power of the timber barons. The provincial government, desperate to avoid an economic crisis, promoted the development of what was to become New Brunswick's next dominant industry: pulp and paper. In return for investing the massive capital required for pulp and paper processing, industry demanded long-term access to Crown forests. The new pulp interests required little from local populations apart from a stable labour supply, and offered little in return apart from employment. But even this benefit to communities began to erode in the 1950s (Parenteau 1994) when local entrepreneurs, community leaders and labour organisations first criticised the major pulp and paper companies for failing to use the public forest in the interests of the people.

The province's forests were wantonly exploited throughout the first half of the twentieth century. Cuts often outstripped growth (New Brunswick Forest Products Association 1944) and timber was usually poorly allocated. From the early 1800s onwards, there was a constant switching from one tree species to another, as the quantity and quality of each successive species declined. As the lumberman faced the possibility of running out of marketable pine, spruce became the next focus of exploitation. As spruce declined and the pulp and paper industry emerged, balsam fir became the staple. This 'highgrading' has continued as trees of smaller and smaller dimensions are cut. Today, increasing volumes of poplar and short-lived hardwoods are being used as a source of fibre.

This shift in trading patterns had a massive effect on forest relations in New Brunswick. The sudden increase in demand favoured the development of larger-scale logging enterprises. By the middle of the nineteenth century, many of the leading timber barons in the area had consolidated their positions by purchasing extensive tracts of land or acquiring long-term leases to high-density woodlands in order to fuel their mills (Wynn 1981). This elicited considerable public outcry and debate on the floor of the provincial legislature, but the concentration of control over New Brunswick's forests continued.

Paradoxically, another persistent trend in New Brunswick forest policy has been the justification of large Crown leases on the grounds that they promote better forest management. In 1982 the Crown Lands and Forests Act eradicated all possibility of local interests gaining access to New Brunswick's forest resources. Justified on the grounds of 'efficiency of wood distribution' and an impending wood supply shortage, the new Act concentrated the previous 84 licences into ten, and required that the licence holders own processing plants in the province (Government of New Brunswick 1980). This entrenched the system of vertical integration that exists in the province's forest industry today, effectively industrialising the Crown forests.

In order to understand the present state of affairs it is important to keep in mind that this monopolisation of licences and vertical integration is historically rooted in the timber barons of the nineteenth century. Now this system is seen by government as the best way to maintain a predictable supply of wood fibre to the 10 pulp and paper mills in the province. The threat of mills shutting down carries considerable political risk: unemployment is already high, and the mills are principal employers in many regions. When an impending wood fibre supply shortage was identified in 1976, despite recommendations to the contrary, the present system was established as a way to keep all the mills running.

A Crisis in New Brunswick's Forests

A crisis in New Brunswick's forests is emerging on three fronts: ecological, social, and economic. Destructive logging practices that continue even today have affected many forest species, together with the structure and dynamics of the forest itself. Shorter-lived, shade-intolerant tree species increasingly dominate New Brunswick's forest composition. The landscape is badly fragmented, with a dense network of logging roads. Short harvest rotations maintain forest ecosystems in a young state. Clearcutting is still by far the most prominent harvesting method on Crown lands (New Brunswick Department of Natural Resources and Energy 1994). Intensive silvicultural practices – including the application of herbicides, thinning and plantation establishment, paid for by public funds – enable the softwood harvest level to exceed the annual rate of growth by 30 per cent. Forestry in New Brunswick is taking on the characteristics of large-scale industrial farming. A Department of Natural Resources and Energy document touts the extent of New Brunswick's 'productive woodland', which covers 97 per cent of the province's forested surface. This figure is twice the national average (Caldwell 1996). Sucked in by what industry and government consider to be a 'fibre supply shortage', every stick of marketable timber in the province's working forest will eventually end up in one of the pulp, veneer or lumber mills.

The effects of 200 years of intensive logging on New Brunswick's wildlife are very evident. The top-level carnivores, the cougar and wolf, have virtually been extirpated from the forest, although a remnant population of cougar may still exist. The Canada lynx (*Felis lynx*) is rare and listed as an endangered species. The woodland caribou was extirpated in the 1930s as the old-growth forests disappeared and whitetail deer (*Odocoileus virginianus*) populations increased (Clowater and Coon 1996). Populations of other species of flora and fauna that depend on older un-fragmented forests are at historical lows in several regions of the province, including the pine marten (*Martes americana*), the fisher (*Martes pennanti*) and a host of vascular plants. Only 1.2 per cent of the province's forests are established as Class I or II protected areas as designated by the International Union for the Conservation of Nature (IUCN). This figure is less than one quarter the national average (Hummel 1995).

The most obvious symptom of the emerging economic crisis in the forest sector lies in the continual drops in employment in the primary (harvesting) sector of the forest industry. Since 1988 employment in the sector nationally has decreased from 500,000 to 360,000 (Statistics Canada 1995). In New Brunswick the workforce of loggers has shrunk from 7,400 in 1981 to 5,690 by 1991. Paradoxically, these declines in employment are not accompanied by drops in timber harvesting or paper production. Indeed, from 1981 to 1992, 'total forest production' levels have risen from 7.8 million to 9.2 million cubic metres (New Brunswick Statistics Agency 1995). This discrepancy is largely due to the increased mechanisation of harvesting and processing, and the increase in the allowable cut based on extensive silviculture.

In the face of these economic and ecological problems, there are no mechanisms for citizens to participate meaningfully in decision making on forest policy. Calls for citizen participation and devolution of power have gone unheeded (CCNB 1994). Goals and objectives for Crown forests are established by the provincial government and are not open to public discussion. While the Department of Natural Resources and Energy will require companies holding licences on Crown land to create 'advisory bodies' in the development of their 1997–2002 management plans, it is doubtful whether these committees will effect anything but superficial changes. In the absence of any meaningful decision-making power, citizens and workers have had to resort to demonstrations and protests to express their dissatisfaction with what is occurring in Crown forests. In 1987, 400 members of the newly formed New Brunswick Woodcutters Association blockaded the main entrance to Consolidated Bathurst's pulp mill (now owned by Stone Container) to protest poor pay and living conditions. In the early 1990s hundreds of wood workers organised demonstrations in the woods to force mechanical harvesters to cease operating. In 1992 thousands of woodlot owners marched through the streets of the provincial capital to protest government plans to undermine their bargaining power for price with the pulp companies. In 1994 students blocked a logging road in an uncut area of the Appalachian Mountains, called the Christmas Mountains, in an effort to see it protected as a wilderness area.

Community Forestry as an Alternative in New Brunswick

In light of the emerging social, economic and ecological crises in the New Brunswick forest sector, it is clear that new ways of governing and managing New Brunswick's forests need to be found. The Conservation Council initiated a campaign in 1993 to reform forest management on New Brunswick's Crown lands. The Council had been established in 1969 by people concerned about environmental issues ranging from the spraying of DDT on the province's forests to loss of topsoil caused by intensive industrial agriculture. The Council's rural development and environment programme promotes sustainable agriculture, ecologically based fisheries management and community forestry. Central to these initiatives is the notion of community-based resource management and a recognition that there are ecological limits to resource use.

The 1993 campaign emphasised a restructuring of the land tenure system to give forest-based communities access to wood on Crown lands, and a devolution of power to give citizens some decision-making power over what happens in the woods. In such a system, provincial forest policy would promote the common good by maintaining and restoring the ecological integrity of New Brunswick's forests and ensuring that forest use is socially responsible. A policy proposal, entitled 'Public Lands in Public Hands: Managing Forests in the Public Interest', published in 1994, was widely circulated. The proposal envisions a system of community forest boards that would have direct control over adjacent Crown lands, whereas more distant Crown lands would be managed in conjunction with the provincial government and the licensee. In its proposal, the Council suggested that community forest boards be composed of elected represen-tatives from rural community committees, Indian communities, woodlot owner organisations, millworkers' unions, woodworker organisations, local conservation and recreation groups, and members at large. To aid this board, a community forester and technicians would be employed.

The proposal was distributed to a variety of forest interests. It was well received, and was endorsed by a number of groups. Petitions containing hundreds of names supporting the proposal were tabled in the provincial legislature. But the government's response to the Council's policy proposal was negative. Bureaucrats and elected representatives justified the current Crown lands management system in terms of its efficiency, and ability to protect the 'public interest' effectively (CCNB correspondence 1993). Nevertheless, communities in British Columbia, Ontario and Quebec are experimenting with the concept. Quebec in particular seems committed to pursuing this approach and is examining a variety of models. The second phase of the campaign thus focused on exposing citizens and interested groups to practical experience with community forestry in New Brunswick and other provinces of Canada.

The Council organised a 'New Brunswick Community Forestry Conference' in 1995 to bring together community forestry practitioners from across the country. By demonstrating the success of the concept in other regions, the Council hoped to help motivate those interested in community forestry and allay the fears of those who oppose this grassroots approach. Participants included representatives of First Nations, municipalities, woodcutters, woodlot owners, economic development corporations, planning commissions, conservation groups, pulp companies, provincial and federal government officials, academics, and unions. Building on the momentum gained at the conference, the Conservation Council initiated a series of local workshops to encourage citizen advocacy for community forestry. These meetings attempted to draw local people who had a direct interest in managing their forests for community benefit.

In its broadest sense, community forestry is forest management conducted by local people, to benefit local people. It is characterised by local control in decision making, and fosters the economic independence of the community. In this way, community forests should be designed to improve local economic, social and environmental conditions. The Food and Agriculture Organisation of the United Nations (1978, p. 19) defined community forestry as 'a set of interconnected actions and works executed by local community residents to improve their own welfare'. Community forestry has three major components: (1) *community empowerment and participation*: community forestry is based on the notion that important resource decisions should be made by those who are directly affected by the consequences of those decisions; (2) *sustainable forestry*: as most rural New Brunswick communities are dependent on forests for economic, ecological and recreational needs, true incentives exist to manage for sustainability; and (3) *community economic development*: community forestry contributes to local economic stability. Unlike the current process, in which money leaves the community to pay distant shareholders, wealth generated from the use of the forest tends to stay in the community. In New Brunswick three communities have adapted the idea of community forestry to fit within current political and social constraints in an effort to meet the needs of their local areas. The following examples demonstrate the potential benefits of community forestry and illustrate some of the current obstacles.

Eel Ground Community Forest

Eel Ground is a Micmac First Nation community of about 300 located southwest of Miramichi City. While the Micmacs have occupied this area of New Brunswick since long before European settlement, the colonial government established this reserve land in 1805 as a permanent settlement for local native people (Hamilton 1984). The Eel Ground forest consists of two major forested tracts totalling 2,832 hectares, both of which are categorised as federal government 'Indian Reserve' lands. The forest is characterised by spruce, fir and pine stands, eastern cedar stands, and poplar and birch stands (Damecour *et al.* 1986). The two forested tracts have been logged at varying levels for fuelwood, pulp and lumber for over a century and a half by both natives and non-natives. The community exercised no control over the use of its forest land and a 'tragedy of the commons' was the result.

During the 1980s, an estimated 20 woodcutters relied entirely on cutting pulpwood for their incomes. The current degraded state of the forest reflects this period of exploitation (Ginnish, personal communication, 1994). In 1986 the Eel Ground Chief and Council decided that this uncontrolled cutting had continued for too long. The Council realised that if timber extraction persisted at the present rate there would be no forest left for future generations. Outside consultants were hired to recommend management plans. An Eel Ground band member with experience in the forest sector took on responsibility for implementation of the plans in 1990. Unrestricted cutting ended, and a process of forest rehabilitation began. Funding was provided by the Canadian Forest Service. Initially, obtaining economic benefit from fibre production was the primary objective of the management plan. The Band adopted stricter controls over timber management to ensure a more secure wood supply and continuous employment (Damecour *et al.* 1986). During the past five years, however, the objectives have expanded to reflect input from the community.

The forest managers at Eel Ground have asserted their belief that in the forest 'everything should be given a fighting chance' (Ginnish personal communication, 1994). In many cases this has meant adopting practices that are much 'softer' than are typically practised on Crown land, for example:

- To preserve wildlife habitats, a series of 'residual areas' has been set aside from harvesting.
- Where possible, a system of management is used which promotes unevenly aged growth.
- To promote native biodiversity, clearcuts are small, and often snags and large unmarketable trees are left standing.
- In accordance with the views of the Council and most community

members, no herbicides or pesticides are used in the Eel Ground forest.

- Tree biomass is left on site to help return valuable nutrients to the soil.

A 1994 survey of the Eel Ground community (Betts 1995) found that the majority of community members agreed with the new approach to the use of their forest. Initially some community elders were concerned about the harvesting of small trees in 'thinning' operations, and the destruction of plants crucial to traditional medicine. As a result, the Band's forest managers have encouraged elders to identify significant and sensitive areas so that these may be set aside.

Heightened access to the Eel Ground forest has increased community use of the area (Denny, personal communication, 1994) for walking, snow-mobiling, and collecting wood for crafts. This increased use has created a demand by more community members to participate in forestry decisions. Unlike several other community forests in Canada, Eel Ground does not have a formal community board charged with making forest management decisions. Informal networks have been the most common means to gain community input. It is reasoned that as the community is relatively small, and everyone knows everyone else, concerns about forest management filter through to decision makers (O'Neil, personal communication, 1994). Steve Ginnish, the Forest Resources Officer at Eel Ground, encourages this sort of feedback. The 1994 survey found, however, that informal networks alone were probably not adequate as a means for community members to participate in management decisions. More than 40 per cent of the people surveyed expressed an interest in participating more directly in forest management at Eel Ground. Nevertheless, alternatives to informal networks might not fit well with the cultural traditions of many bands. Before opting for change toward non-native community forest structures, band governments and aiding organisations must consider alternatives that are culturally appropriate.

One of the major benefits of the Eel Ground community forest has been the relatively stable employment provided for between 10 and 17 silvicultural workers. Rather than becoming a 'make work' project, where wood workers are provided with limited funding so as not to disqualify them for unemployment benefits, the forest managers have ensured that employment is provided for at least eleven months of the year (Ginnish, personal communication, 1994). While substantial funds have been provided by the Canadian government to finance the silvicultural work, each year the operation moves closer to self-sufficiency. Employment will undoubtedly increase as plans for the production of diversified wood products come into effect. In the autumn of 1995 Eel Ground purchased a portable sawmill so that it has the capability to make lumber for house construction and flooring, and shingles for roofs.

Still, the economic potential of the Eel Ground forest is constrained by the limited land base and the degraded state of the forest. Attempts to gain access to Crown land so that the Eel Ground community forest would be more viable have failed. The forestry operation itself remains non-profit. Any revenue from harvesting either goes back to pay for further management, or is put in a trust fund for special community events. Fuelwood is donated to community facilities, including the school, the band hall and the rehabilitation centre (Ginnish, personal communication, 1994).

Moncton Community Forest (Irishtown Nature Park)

Unlike the Eel Ground community forest, the primary objective of the Irishtown forest is not timber production but recreation and education, though some timber extraction does occur. The Irishtown Forest is located about a kilometre north of the city of Moncton in southeastern New Brunswick. The forest has a great diversity of trees by age class and species type. Stands vary from predominantly balsam fir (*Abies balsamea*) and black spruce, to mixed softwood (spruce, fir and tamarack – *Larix laricina*) and hardwood (maple and poplar) (South Eastern New Brunswick Wood Marketing Board 1992). This 712-hectare forested area had served as a drinking water reservoir for the city since the early 1800s. No longer necessary for drinking water requirements, the forest is now partially managed by a community board called the Irishtown Nature Park Committee.

Forest management at the Irishtown forest has evolved since 1993 when the area was first designated as a recreational area. Initially the city of Moncton had plans for extensive timber harvesting. But once the city's Community Services Department realised that the forest's unique features made it ideal for wildlife habitat and recreation, these plans were altered. The Department's call for input from recreational groups in the city met with an enthusiastic response. As a result the Irishtown Nature Park Committee was established and given the authority to plan the future use of the forest. Many of the member groups of the community committee rejected high-impact harvesting techniques such as clearcutting. Others felt that some 'over-mature' areas should be left for wildlife habitat. While harvesting was not terminated, it was decided that a much more cautious approach would be adopted. Maintenance of wildlife habitat and nature interpretation skills have become central goals. A citizen-based monitoring programme has been established so that committee members and the groups they represent can serve as watchdogs over timber management at Irishtown. The broad interests involved in the use of this forest have ensured that the area is managed for a wide range of objectives that go beyond timber production.

Currently the Irishtown Nature Park Committee consists of nine community group representatives, a community forester, a member of the Community Services Department, and a city councillor. The committee is constantly soliciting the participation of new groups. The relationship of the Committee with the city of Moncton is still evolving. Interviews with committee members revealed that perhaps they have less control over actual timber management than they do in planning for the recreational development of the forest. The city continues to own the land and provides financial and logistical support to the community forest.

The Irishtown Nature Park is the only extensive natural area that exists within the Moncton city limits. Already local schools are taking advantage of the area to teach children about forest ecosystems, ecological processes, and wildlife. Such environmental education is crucial in an urban setting, and contributes to what forester Aldo Leopold termed a 'land ethic'. If people begin to understand and appreciate their natural environment, they will have more incentive to be good stewards. Wood harvesting in the Irishtown forest has generated tens of thousands of dollars in revenue over the past two years, which is to be reinvested in the forest (Hawker, personal communication, 1994). Only local contractors were hired to harvest the timber.

The Kedgwick Loggers' Co-op

A third example of community forestry in New Brunswick exists near Kedgwick in northeastern New Brunswick. Here relatively small tracts of Crown land are managed by the Kedgwick Loggers' Co-op. When the local mill was closed down, 16 separate parcels of Crown land were no longer allocated for harvesting. As the existing licensee felt that forest blocks were too small and isolated to make harvesting economically viable, the provincial government intended to sell them off to the highest bidder. The newly formed Loggers' Co-op proposed to government that they produce management plans for the land and take on management responsibilities. Conscious that it might set a precedent by entering into an agreement directly with the Co-op, the provincial government suggested the loggers pursue an agreement with the pulp company. An agreement was reached, with the proviso that the timber extracted from the area would go to the pulp company.

The Co-op uses a variety of cutting methods, none of which is highly mechanised. In stream buffer areas, for example, horses are used to haul logs to roadside. In addition to providing livelihoods to Co-op members, profits from harvesting go toward supporting the wages of local workers who maintain and monitor the environmental quality of the nearby Restigouche River and to other community initiatives (Guerette, personal communication, 1995).

Since the current arrangement dictates where the timber must be processed, the Co-op is constrained in terms of developing more profitable markets and value-added products. The loggers hope, however, that responsible management of these isolated blocks of Crown forest, and the obvious community benefits derived, will bolster their case for more autonomy in determining how timber is utilised.

The Potential for Community Forestry in New Brunswick

While the Eel Ground, Moncton, and Kedgwick community forest initiatives are isolated cases, they hint at the potential for this decentralised approach to benefit local communities and sustain or restore the health of forest ecosystems. Nevertheless, many obstacles to expanding community forestry remain. Most First Nation communities have access to only tiny amounts of land, which is often of poor quality. The Maliseet and Micmac are relegated to reserves that range from 3,907 hectares to as small as 26.4 hectares. Most of these are not large enough to support viable timber businesses. In New Brunswick, municipalities typically own very small forests. Where some potential exists, as in the case of Saint John, little effort is made to involve community members in forest planning and timber management. Forests are seen simply as an additional source of municipal revenue. Crown forests, meanwhile, which cover some 3.1 million hectares of land, are locked up by multinational pulp and paper corporations as a result of the Crown Lands and Forest Act.

If native communities, municipalities and the hundreds of forest-based communities in the province are to have a chance to pursue community forestry, the government legislation which allocates all of New Brunswick's Crown land to wood-processing companies must be changed. This will not be an easy task. New Brunswick's Crown Lands and Forests Act has been touted by those in power as Canada's best forest management system (*Telegraph Journal*, 1 April 1995). It guarantees the pulp companies secure access to massive tracts of forest at less than the market price for private wood. It enables them to maximise the timber extraction on that land, using highly mechanised logging practices, and it permits them to manage primarily for fibre production. It also provides for the reimbursement of their silvicultural expenses such as herbiciding, tree planting and thinning. For these reasons large industrial forest companies have a strong interest in maintaining the *status quo*.

Provincial government decision makers also appear afraid to change the current 'efficient' system (Hatheway, personal communication, 1994). Given New Brunswick's overwhelming economic dependence on the forest sector, the Department of Nature Resources and Energy is reluctant to adopt a concept that remains relatively untested in the province. In particular, the government is concerned that if community forestry means that communities expand the objectives of forestry in their local forests beyond maximising fibre production, the flow of fibre to the ten pulp and paper mills will be reduced.

Other hindrances to the proliferation of community forestry have more to do with the political and social culture of New Brunswick than government policy. As the local pulp mill is often the primary employer in many towns, people are frightened that criticism of the *status quo* or advocacy of an alternative approach to forestry could get them or members of their family fired, or destroy future job opportunities for family members (woodcutter interviews, 1994). Critics that do emerge in local communities are sometimes socially marginalised.

In spite of these obstacles, there are several trends internationally, nationally and provincially which may open some political and social space for experimentation with community forestry. European market pressures for sustainably produced wood products are being felt keenly by industry and government. Green labelling and certification of forest companies has the potential to force changes in forest policy and practices. The Canadian Standards Association's draft certification standards, for example, will require meaningful public participation in forest decision making.

At the national level, the continued movement by First Nations toward self-government may speed the adoption of community forestry in native communities, particularly if land claims secure access to Crown forest lands. At the provincial level, a community-based approach to rural land-use planning is being implemented. While it explicitly excludes Crown land-use planning, once local people gain experience with local planning this may create a demand for control over Crown land. Its institutional framework of rural community committees and regional planning commissions could well accommodate community forest boards.

The current drive to cut government spending may also provide an incentive to adopt community forestry. In an effort to cut the costs of bureaucracy, the Quebec provincial government has recently begun to decentralise forest management responsibility to municipalities in the Abitibi-Temiscaming region (Masse 1995). If this experiment succeeds, it may encourage New Brunswick to attain savings through similar approaches. In the light of the current social and political climate, however, it is more probable in the short term that limited forms of community forestry may be pursued by local groups when special circumstances arise. Government policy will only change to permit communities to have control over forests in the face of a concerted demand from a variety of interests.

The Road to Community Forestry in New Brunswick

Unfortunately, given the context in which citizens of New Brunswick's forest-based communities work and live, a massive groundswell of public demand for community forestry is unlikely in the immediate future. Too many rural citizens and small business owners are fearful that taking a public stand will jeopardise their economic future. Too many people in urban centres have been led to believe that Crown forests are being managed for the common good, and that the ecological integrity of our forests and the future of rural communities are not in danger. In this way, the emerging crisis is perhaps more insidious for its subtlety.

Pressure for more local control over Crown forests must develop and gain sufficient strength before the timber supply runs short, mills shut down, and communities are left with simplified, degraded forests. But the ranks of those prepared to push for change are growing. These include woodlot owners who face unfair competition from timber extracted from Crown lands; woodcutters displaced by mechanical harvesters; those locked perpetually in cycles of unemployment; speciality wood product manufacturers who have no access to Crown timber; trappers, hunters and naturalists who see habitats for many forest species disappearing; and outfitters and recreationalists who see the quality of outdoor experiences diminishing.

Alone, the Conservation Council will not persuade government to experiment with community forestry. The best functions it can perform are to build broad public support for community forestry; to facilitate discussions among and within different communities and groups interested in the idea; to act as a clearing-house for information; and to continue to research into models of local resource management. Although external forces – such as foreign markets, certification, funding agencies and various levels of government could facilitate experimentation with community forestry, if it is to be successful the initiative must come from the bottom up, rather than from the top down.

Several actions can be taken which will make New Brunswick's political, social and economic environment more conducive to this change. First, there must be a fundamental change to the Crown Lands and Forests Act. The policy must be made more flexible so that innovative community forestry experiments are allowed to go forward. Second, funding that is currently applied to forest planning and timber management carried out by the federal government must be made available to those communities enthusiastic about initiating community forests. External funding plays an extremely important role, particularly in the early stages of establishing a community forest project. Every year, millions of dollars flow from the public purse to the private sector to support silviculture on Crown land

and improvements in pulp and paper processing. It would be appropriate to redirect some of this public money to community forest initiatives.

Finally, the dissemination of information, and education on forestry issues must continue. Only a well-informed and enfranchised public will be concerned enough to see the flaws in the present system and demand alternatives. Unfortunately, the level of alienation that exists between citizens and their governments in Canada is extreme. Governments are increasingly seen as serving clients with vested economic interests rather than the public good. The public reaction is to admonish government, demand less of it, or abrogate their democratic responsibilities altogether. The more this occurs the more isolated government becomes from its citizenry, and the more it finds itself catering to special interests. John Ralston Saul (1995: 76) wrote: 'People become so obsessed by hating government that they forget it is meant to be their government and is the only powerful public force they have purchase on.' It is time to take advantage of this public force and demand management that is in the interest of our communities and our forests.

Postscript

With regards to hunting, trapping and fishing, Indians are not restricted to reserve lands, as treaties and court decisions affirm their right to hunt and fish virtually anywhere. Moreover, a recent court decision in New Brunswick has affirmed this right for the commercial harvest of timber from Crown lands. This decision – which, in essence, opens up Crown lands to commercial harvesting by natives – is being appealed by the province. Meanwhile, native communities are looking into the possibilities of native management or co-management arrangements with the government on Crown land.

BIBLIOGRAPHY

Betts, M. (1995) 'An Analysis of the Potential for Community Forestry in New Brunswick', Faculty of Environmental Studies, University of Waterloo, unpublished Masters thesis.

Caldwell, R. (1996) 'New Brunswick: You Can See the Forests for the Trees', New Brunswick Department of Natural Resources and Energy Home Page.

Clowater, R. and Coon, D. (1996) *Impoverishing Nature: the Biodiversity Crisis*, Fredericton: Conservation Council of New Brunswick, Fredericton.

Commission on Land Use and the Rural Environment (CLURE) (1993) *Final Report*, Fredericton: Minister of Supply and Services.

Conservation Council of New Brunswick (CCNB) (1994) *Public Lands in Public Hands: Managing Crown Forests in the Public Interest*, Policy Proposal, Fredericton.

Damecour, G., Chase, B. and Frittenberg, K. (1986) 'Eel Ground Reserve No. 2 Forest Management Plan', unpublished plan, Eel Ground.

Dunster, J. (1991) 'Community Forestry: What is it?', in: P. Smith and G. Whitmore, (eds)

Community Forestry, Proceedings of the Lakehead University Forestry Association. 23rd Annual Symposium.

Food and Agriculture Organisation of the United Nations (1978) *Forestry for Local Community Development*, FAO, Rome.

Forest Products Commission (1994) *The Directory of New Brunswick Forestry Sector Groups*, New Brunswick Forest Products Commission, Fredericton.

Government of New Brunswick (1980) *The Crown Lands and Forests Act*.

Hamilton, W. D. (1984) *The Julian Tribe*, The Micmac-Maliseet Institute, Fredericton.

Hummel, M. (1995) *Protecting Canada's Endangered Spaces: an Owner's Manual*, Key Porter Books, Toronto.

Leopold, A. (1966) *A Sand County Almanac*, Oxford University Press, Oxford.

Lower, A. R. M. (1973) *Great Britain's Woodyard: British America and the Timber Trade 1763–1867*, McGill-Queen's University Press, Montreal.

Manley, S. (1989) Transcribed speech from the meeting of the Forest Committee of the Environmental Coalition of PEI and the Eastern Woodlands Association, 3 November.

Masse, S. (1995) *Community Forestry: Concept, Applications and Issue*, Minister of Supply and Services, Quebec.

Ministry of Supply and Services (1994) *The Canadian Pulp and Paper Industry: a Focus on Human Resources*, Price Waterhouse, Toronto.

Mladenoff, D. and Pastor, J. (1993) 'Sustainable Forest Ecosystems in the Northern Hardwood and Conifer Forest Region: Concepts and Management' in: Aplet *et al.* (eds) *Defining Sustainable Forestry*, Island Press, Washington.

New Brunswick Department of Natural Resources and Energy (1994) *Forest Management Plan Highlights Crown Land, License 1–10*, Minister of Supply and Services, Fredericton.

New Brunswick Forest Products Association (1944) *Forestry and Post-War Reconstruction in New Brunswick*, brief presented to the New Brunswick Committee on Reconstruction, 24 January.

New Brunswick Reader (1996) 'Timber Land', 17 February 1996, pp. 4–20.

New Brunswick Statistics Agency (1995) *The New Brunswick Economy: a Report to the New Brunswick Assembly*, Queen's Printer for New Brunswick, Fredericton.

Parenteau, W. (1994) 'Forest and Society in New Brunswick: The Political Economy of Forest Industries 1918–1939', University of New Brunswick, Fredericton, unpublished PhD thesis.

Perley, M., (1847) 'Report on the Forest Trees of New Brunswick', *Simmons' Colonial Magazine*. Vol. 9, No. 42 (June), pp. 129–55.

Sandberg, L. A. (1992) 'Introduction: Dependent Development and Client States: Forest Policy and Social Conflict in Nova Scotia and New Brunswick', in: L. A. Sandberg, (ed.), *Trouble in the Woods: Forest Policy and Social Conflict in Nova Scotia and New Brunswick*, Acadiensis Press, Fredericton.

Saul, J. R. (1995) *The Unconscious Civilisation*, Massey Lecture Series, Anansi Press Ltd, Concord, Ontario.

South Eastern New Brunswick Wood Marketing Board (1992) 'Woodlot Management Recommendations for the City of Moncton', unpublished.

Statistics Canada (1995) *Annual Estimates of Employment, Earnings and Hours 1983–1994*. Minister of Industry, Science and Technology, Ottawa.

Wynn, G. (1981) *Timber Colony: a Historical Geography of Early Nineteenth Century New Brunswick*, University of Toronto Press, Toronto.

Newspaper articles

Daily Gleaner, 'Fight to Continue to Stop Cutting of Christmas Mountains Forest', 20 July 1993.

Telegraph Journal, 'Inquiry Told Woodcutters Are "Treated Like Animals"', 18 July 1993.
— 'Jobless woodcutters protest use of mechanical harvesters', 15 April 1994.
— 'Axe hovers over Christmas mountains', 22 January 1995.

Interviews

Denny, James. Band Councillor, Eel Ground Reserve, New Brunswick, 5 November 1994.

Ginnish, Steve. Forest Development Officer, Eel Ground Reserve, New Brunswick, 15 October 1994.

Guerette, Martin. Kedgwick Loggers Co-op, Kedgwick, New Brunswick, 1 November 1995.

Hatheway, Harold. Canadian Forest Association, Fredericton, New Brunswick, 6 December 1994.

Hawker, Heather. Forest Manager, City of Moncton, New Brunswick, 19 October 1994.

O'Neil, Jane. Community Development Officer, Eel Ground Reserve, New Brunswick, 6 November 1994.

Woodcutter interviews conducted in Mirimachi during the summer of 1994. Interviewees were guaranteed confidentiality.

14 Canadian Hunters Fight for the Forest

The Algonquins Striving for Territory and Good Management

BOYCE RICHARDSON

in collaboration with

RUSSEL DIABO

In the fall of 1988, a group of Algonquin Indians arrived in the Canadian capital, Ottawa, and, as a form of protest, set up their tents on the grounds of the national Parliament. Today, after more than seven years of continuous struggle, the band has been given a decisive voice in designing and implementing a sustainable forestry model on their traditional land. The Algonquins' right to a viable life within the modern Canadian economy is finally respected.

When the battle started in 1988, the Algonquins had come to Ottawa from their home in a rolling, forested wilderness 300 kilometres miles north of the city, where they have lived since long before Europeans arrived in Canada. Initially the Algonquins had adapted to the small-scale logging, hydroelectric dams, sports hunting and tourism that came into their lands. But with the beginning of mechanised clearcut logging in the 1960s they found their way of life slowly being hacked to pieces by commercial forestry companies, and by 1988 the pressures had become intolerable. Although they had always been a quiescent, uncomplaining people, now they had no choice but to emerge from their forest in a dramatic public protest against the destruction of their way of life.

What they were trying to do was extremely interesting, and, at the time, unprecedented in Canada. They knew that in 1987 a UN Commission on global environment and development had reported to the UN General Assembly that the Earth is in serious danger unless we can create a new, environmentally sustainable form of development. Even more to the point, they knew that the Commission had praised the knowledge and wisdom accumulated by the world's indigenous peoples, and had recommended that aboriginals be given 'a decisive voice' in all decisions being made about development of their traditional lands. And they knew one other thing: the federal, and every provincial government in Canada had endorsed the report, and the Prime Minister, Brian Mulroney, had been acclaimed internationally for his support of it. Now the Algonquins metaphorically waved the report in Mulroney's face and said it was time for him to put up or shut up: they wanted 'a decisive voice' in decisions about their lands. The government's response was to arrest the Algonquins and charge them with committing a nuisance on Parliament Hill.

This Algonquin group – about 450 people strong today – is one of ten small communities which have always been known as Mitchikanibikongink – People of the Stone Weir – because centuries ago they settled around a natural stone bridge over which the headwaters of the Ottawa and Gatineau rivers would flow together at certain times of the year. The Algonquin hunters, like other Indians in the huge expanse of the Canadian boreal forest, have always depended for their food on the animals of the forest – moose, the most important, but also marten, beaver, lynx, bear, a variety of birds such as geese, ducks and ptarmigan, and many species of fish, notably walleye, pike, trout and sturgeon.

Over the 6,000 years that they have inhabited this forest, they have achieved an extraordinary knowledge of its biology and of the behaviour of animals. But their knowledge has always been treated by governments and invading Canadians as irrelevant and useless in the economy of a modern nation. Logging of their forest began in 1870. In 1871, the Stone Weir was flooded by a logging company's dam. By 1928 a huge hydroelectric reservoir had been built in their hunting lands. In the same year the Quebec government, without any consultation, created a so-called wildlife reserve in one million hectares of their traditional hunting territory. Since then, that reserve has been handed over to private companies for logging, and to tourist outfitters for sports hunting.

By the time the Algonquins appeared on Parliament Hill, almost half of their hunting territory had already been clearcut, and the patrons of 12 tourist lodges were taking the animals the Algonquins lived on, all of this with devastating impact on their physical, emotional and spiritual health. When a new forest management plan was adopted by the government in 1985, aboriginal use of the forest was not even mentioned. Soon after, a government agency was established whose purpose was to make money out of sports tourism. All of these measures conflicted with traditional Algonquin use of the forest and repeated a well-worn historical pattern – in more than 100 regulations passed by Quebec governing the wildlife reserve over a period of sixty years, the Algonquins were mentioned only once, and that was to say that their rights within the reserve no longer existed! That the Algonquin protest on Parliament Hill in 1988 should be virtually ignored, then, was simply consistent with Canada's long, historical violation of aboriginal rights in the land.

The result of this policy is clear to anyone who visits the impoverished Barriere Lake reserve. A study made in the late 1980s indicated massive unemployment, extensive dependence on alcohol, and a complete absence of the management skills the Algonquins would need if they were to gain control over their own affairs. Thirty per cent of the people were under sixteen, and the population was expected to double to more than 900 in the coming fifteen years. Yet there was no room for more houses on the 23 hectares which is all the land granted them by Quebec as their reserve.

In making their protest, the Algonquins had to confront not only the enormous weight of history, but also the assumptions underlying all government policy towards development. Clearcut logging treats the forest as if it were simply a source of timber, and pays no attention to animal habitat, water retention, stream integrity, soil cover, fish spawning, beaver streams, or aboriginal use. In this industrial mind-set, forest is timber, trees are wood, and nothing else.

The Algonquin proposal, however, was quite unusual and extremely creative. They had decided to press the federal and provincial governments to join with them in creating a conservation strategy covering their traditional lands, a strategy that would not only provide for the interests of loggers, sports hunters, and tourist outfitters, but also for their own traditional use of the forest. By implication they had to bring indiscriminate clearcutting to a halt.

For almost a year after their appearance on Parliament Hill, governments refused to talk seriously with the Algonquins about their hopes for the future. But then occurred the spark that ignited their real anger. On 28 August 1989, one of the Barriere Lake people was told by workers from the Quebec Ministry of Energy and Resources that he had to move out of his home because they were about to spray the forest with a chemical called glyphosate. Something similar had happened in the previous summer, when some Algonquins had become violently ill after eating berries that, unknown to them, had been sprayed. As a matter of urgency, the Algonquins now tried to set up a meeting with the provincial ministry of forests, but were refused. 'We rely heavily on the fruits of this land for our subsistence,' stated their Chief Jean-Maurice Matchewan, in a letter to the minister. 'Your government knows this, yet nobody asks us if it is all right to spray there. Having been denied what we feel is a reasonable request, we have no choice but to take the matter into our own hands. This will advise that we have set up a blockade to stop ... chemical spraying within our traditional lands.'

This blockade brought out most of the community in enthusiastic protest, and they won their point: the government reluctantly agreed to postpone further spraying until the following year. Pressing their advantage, the Algonquins tried to set up further meetings, but got nowhere. They began to realise that the spraying was merely a symptom of the whole problem, which was the determination of government and industry to press on with clearcut logging in defiance of the Algonquins' clearly expressed opposition. They then began a series of blockades of logging roads which lasted, on and off, for the next two years. They had to defy court injunctions, police harassment and arrest, the pressure of trans-national industry, and deplorable government duplicity. Finally, against all the odds, in August 1991 the federal and Quebec governments signed an agreement, since known as the Trilateral Agreement, to collaborate with the Algonquins in designing an integrated management plan for their traditional lands.

The Agreement provided for a four-year study and inventory of all natural resources over an area of 10,000 square kilometres, and laid down a formal process and timetable. The work was to be done through Special Representatives nominated by each party, who had first to draw up a set of 'provisional measures' for protecting sensitive areas of forest, and then work towards an agreed management plan that would become the object of a negotiation between Quebec and the Algonquins in the autumn of 1994. The one glaring weakness of the Agreement was that the provincial govern-ment's management deals with logging companies were allowed to stand, guaranteeing the companies continued access to the trees. A second weakness turned out to be the reluctance – indeed, the refusal – of the Quebec Ministry of Forests to honour the cooperative nature of what had been signed. From the beginning, this ministry acted as if the right to cut trees would continue to override all other rights. Within a month or two the Agreement had virtually broken down. But the Algonquins continued to map the sensitive areas, using their own minimal funds.

By August 1992 a judicial mediator had to be called in. He found that the provincial agreements signed with the logging companies 'do not respect the Trilateral Agreement, either in the spirit, or the letter'. He said the promised government funding had been withheld, and only the Algonquins had kept the Agreement alive. 'It is David, and not Goliath, who is attempting to sustain the agreement', commented the judge. He described the Agreement as a marvellous tool for the intelligent management of a forest. 'Why then', he asked, 'are we at a point where we can almost see such a beautiful project collapsing?' This did get the Agreement functioning again, but by February 1993 Quebec had withdrawn unilaterally, accusing the Algonquins of being 'unreasonable'. Finally, to cut a long story short, the Algonquins managed to sustain enough pressure to ensure that the administration of the Agreement was taken out of the hands of the Ministry of Forests, and placed directly under the office of the provincial Premier. Two valuable years had been lost, but since then enormous progress has been made – and, largely because of the dogged Algonquin persistence, the atmosphere between the parties has improved vastly. The recalcitrant Quebec government has finally become more co-operative; industry people, who at first regarded the Algonquin claims as a bizarre nuisance, have begun to understand the need to respect the ancient traditions and knowledge of the forest people, and to recognise their right to a viable life within the modern Canadian economy.

The mapping of the sensitive zones has been completed, and for some time now logging crews have been accompanied by Algonquin monitors who have been able to ensure that these areas are not disturbed. These include a wide variety of sites of importance to Algonquins and the animals they depend on, such as moose yards, bear dens, fish spawning sites, beaver streams, sugar (maple) bushes, speciality wood areas, eagle nests, travel routes, and various special sites such as burial grounds, sacred places and old settlements. This information has been collected through intensive interviews with knowledgeable Algonquin hunters. These sites amount to about 12 per cent of the entire area under review, or 1,200 square kilometres.

At the same time work has been done on what the Agreement calls 'measures to harmonise' Algonquin life with the economic activities of outsiders. In both of these areas traditional Algonquin environmental knowledge has, for the first time in Canada's history, been taken into account and integrated into recommendations for future action. All the problems, of course, are not yet solved. The Quebec government still appears to consider the flow of education to be a one-way street – from the European sensibility to the Algonquins. They have not yet quite understood that Euro-Canadians have much to learn from the people who have lived in the forest for thousands of years. Quebec also is reluctant to accept that they are engaged in a process which, if it works, will lead to a co-management regime with the aboriginals covering an immense tract of land over which provincial authorities have for more than a century exercised unrestricted control. More than any other province in Canada, Quebec is jealous of its jurisdiction, and reluctant to share it. The Algonquins' task has been complicated by the recent election of a Partie Quebecois provincial government that has pledged to establish Quebec as an independent nation.

Even if everything proceeds smoothly from this point, the Algonquins face the daunting task of preparing themselves to play a full role in co-management of their lands. They aspire not only to protect the animals and defend their traditional way of life, to improve the habitat and restore its ecological health, but also to upgrade their education and skills so that they can play a full role in everything that happens in their lands. Guiding, silviculture, fisheries development, law enforcement and tourism are just some of the areas for which their special skills in the bush fit them, but from which they have been excluded in the past. Intensive consultations held within the community have identified many obstacles that will not easily or quickly be removed, such as the customary exclusion of aboriginals from maintenance work in the wildlife reserve, their exclusion from decision making and their lack of access to capital to establish small businesses.

The Algonquins, however, know what they have to do to ensure the revival of their damaged culture. Everything depends on regaining control over their own decisions so that self-esteem, confidence and pride in their aboriginal culture can be restored. They are still clinging to the hope, given them by the UN report, of 'a decisive voice' in all decisions affecting them. In asserting their own needs and rights, the Algonquins will inevitably require external support. It remains to be seen how much they will receive. In many parts of Canada sports hunters are already resisting the curtailment of their privileges that usually follows the restoration of violated aboriginal rights. It matters nothing to these sportsmen that they take twice as many moose as the Barriere Lake Algonquins, who depend on the moose for a considerable part of their diet. The Algonquins are aware that outsiders will have to be educated to respect Algonquin traditions and knowledge and to accept their guidance in protecting the forest. As they move to assert their rights, the Algonquins, like other aboriginals in Canada, recognise that they confront systemic and institutional racism, entrenched in the behaviour of a wide range of Canadian institutions. None of this is going to be easy to achieve, but it is amazing, and encouraging, that this small, depressed community has already been able to pressure two powerful governments into undertaking a far-sighted movement towards creation of an environmentally sustainable economy over a huge area of land. That, in itself, is already an immense victory.

Three Scenarios for Future Sustainable Management

The Algonquins are currently at the point of having to produce a viable plan – environmentally and economically balanced – for the sustainable forest management of the Trilateral Agreement territory. The Trilateral Agreement was officially extended until December 1996, with a deadline to table an integrated resource management plan (IRMP) by 31 March 1996. There are potentially hundreds, perhaps thousands, of plausible combinations of resource management alternatives for the Trilateral Agreement territory, reflecting the variety of forest stand composition, soils, resource values, and ways in which resource management interventions could be conducted or impacts mitigated. Prior to the conceptualisation of the resource management scenarios three basic questions must be addressed: (1) What do we want to sustain and why? (2) What goals do we want to attain through the implementation of the IRMP? (3) What values underpin these goals?

At the last meeting of the Special Representatives it was proposed that three broad management alternatives be developed and assessed: (1) *Status quo* scenario; (2) Modified operations; (3) Substantially modified operations, with the application of landscape ecology principles.

All three management scenarios developed must include the required forest sustainability criteria related to timber and non-timber objectives, with associated targets. Each management alternative must be analysed to identify the future forest condition which is expected to result from its implementation, and to assess its ability to ensure natural resource sustainability while producing the desired objectives over time.

A set of screening criteria must be developed and agreed upon by the Special Representatives, which specifies the minimum standards to be met by the three proposed scenarios. These minimum standards must include 'specified' non-timber objectives and targets which are kept constant within each scenario, thereby assuring the sustainability of non-timber values in the selected management alternative.

'Status Quo' *alternative*

This scenario is the baseline from which the acceptability of the other scenarios is to be evaluated. With the development of the successional keys, and information gleaned from sources such as company operating plans, it should be possible to forecast the future condition of the forests of the Agreement Territory and determine through the use of criteria/ indicators whether sustainability objectives are being satisfied under the current management regime.

'Modified Operations' *alternative*

The 'modified operations' scenario may attempt to achieve significant modifications of resource decision making and timber management activities. The basic objective of this scenario, for example, could be 'to maintain the Trilateral Agreement territory as a healthy ecosystem and management of resources for multiple uses, including the maintenance or enhancement of Algonquin traditional land uses and culture'. Differences in the management approach from the 'null' scenario could include, for example;

- Reductions in annual allowable cut/alternative sourcing for wood supply;
- Imposition of harvest exclusion zones (cultural zones, old forests, etc.);
- Imposition of a revised riparian zone management strategy;
- Elimination of ineffectual silvicultural practices;
- Mixed planting of compatible species on selected sites;
- Alternative rotation ages for portions of some working groups;

- Maintenance of protected areas as ecosystem reserves;
- Application of silvicultural systems to promote/encourage natural regeneration;
- Application of landscape ecology principles;
- Imposition of harvest pattern that accounts for and accommodates the geographic distribution of Algonquin resource users;
- Application of landscape visibility analysis strategies in areas of high tourism and cultural values.

'Substantially modified operations' alternative

The third alternative is to be more far-reaching and innovative and is likely to incorporate landscape ecology principles as the underpinning of the management strategy. Management alternatives proposed under this scenario will shift from managing forest states to managing processes – from focusing on trees or stands of trees to focusing on ecosystems. This paradigm shift to landscape management will require viewing management objectives at larger spatial and temporal scales as well as the accommodation of a broader range of objectives. This alternative may incorporate the following:

- Integration of biodiversity maintenance objectives with harvesting;
- Shift in emphasis from maintaining forests in a given state to maintaining particular processes;
- Development and application of modified silvicultural tools to meet a broader set of objectives than timber management criteria – significantly longer rotations, retention of trees, etc.

Update: Domtar Attempting Corporate Takeover of Algonquin First Nation

At the present critical stage of the Trilateral Agreement negotiations, the Algonquins of Barriere Lake are facing an attempt by a dissident group to overthrow the traditional leadership of the community. The dissident group, although in Quebec, is using the Winnipeg law firm of Thompson, Dorfman, Sweatman, which also represents Domtar Inc., a logging company operating on the traditional lands of the Algonquins of Barriere Lake. Domtar is controlled by the Government of Quebec. Domtar Inc. operates within the Trilateral Agreement territory through an entity called Produits Forestier Domtar, which holds the largest Timber Supply Contract (CAAF) within the Trilateral Agreement territory. The IRMP is expected to result in dramatic reductions in Domtar's cutting volumes.

On 23 January, the federal Department of Indian Affairs announced its decision to recognise the (dissident) Interim Council as the 'legitimate Council of the Barriere Lake Band' rather than the Traditional Council of

Barriere Lake. If this order were to be granted by the Federal Court it would entitle Domtar, through their legal counsel, to gain access to the confidential records of the Algonquins of Barriere Lake during this critical phase of negotiations and compromise their rights and interests permanently. It would be just a matter of time before Domtar hijacked the Trilateral process. The people of the Barriere Lake Community have officially rejected the decision.

The Interim Council was formed on the basis of a petition containing 156 signatures plus 12 forged signatures. The names include 50 resident dissident band members, 7 band members who live outside of Canada, 46 Bill C-31 (C-31 recognises native people off reserve) who are non-resident and 9 other First Nation members. The remainder are not resident. The 'recognition' of the Interim Council has already meant that the Algonquins have been cut off from their funds and telephone service, which has left community services, including the school, in total disarray.

Postscript[1]

On 21 October 1996 the Algonquins of Barriere Lake went on road blockade to prevent Domtar and other logging companies from reaching large portions of the Algonquin territory which they intend to clearcut. The blockade was the result of a decision taken by the Ministry of Indian Affairs about 10 months ago to withdraw its recognition from the Algonquin administration. This meant the end of a long and very promising process under the Trilateral Agreement between the Algonquins and the governments of Canada and Quebec, aiming at developing and implementing sustainable management of the traditional Algonquin territory. By recognising a dissident group as the legitimate leadership of the Algonquins, the government in fact abandoned the Trilateral Agreement, thereby opening the road once again to clearcutting of Algonquin land.

The Algonquins seek to solve the conflict over leadership by codifying their customs regarding leadership. They are now urging the government to respect the outcome of this process, and thus the will of the majority of the band. This would make it possible to restore the process under the Trilateral Agreement.

Since mid-January 1997 the Algonquins have prevented all forest companies from entering Algonquin territory. The situation is expected to reach a climax soon. By early February 600 to 1,000 company people may be out of a job if logging of the Algonquin territory does not commence. This is a serious problem, especially in this part of the country which is already prone to chronic unemployment. However, logging of Algonquin land will offer no long-lasting remedy against unemployment. Moreover, the fact that many local people already lost their jobs many years ago due to dismal forestry policies is hardly ever mentioned (personal comments by Joel Damay).

The government governs by crisis. The issue is now on the agenda of talks between the Prime Minister of Canada and the Prime Minister of Quebec. It is to be hoped that choices are made which bring social peace and which enable long-term management.

NOTE

1   By Russel Diabo (1996:7) *Taiga News*, December 1996, No. 19.

15 Finding Common Ground

Adapting Forest Management in Krakow to the Pressures of Modernity

JERZY SAWICKI • BOGUMILA KUKLIK

KATARZYNA TERLECKA• TOMASZ TERLECKI • RAFAL SERAFIN

National Parks Unit, Polish Ecological Club, Poland

The National Parks Unit of the Polish Ecological Club, an environmental umbrella organisation, has started a series of case studies on large forest complexes in Poland. It aims to present government institutions and Parliament with suggestions for policy reform. Noting Poland's agreement to the Forest Principles which were adapted at the Rio Summit in 1992, this chapter draws attention to the impact of the far-reaching economic, legal and institutional reforms currently under way in the Polish forestry sector. There is reason to fear that further privatisation of forests and forest management will go unchecked in the absence of adequate regulations and control from government.

The chapter describes legal and policy issues surrounding the protection of Niepolomnice Forest, located near the industrial centre of Krakow. A major problem hampering adequate management is the conflict between forest and nature conservation authorities which results from an incomplete division of jurisdictional responsibilities. In addition, the forest suffers from serious insect pests, fungal diseases and other damage due to the alteration of drainage patterns and air pollution. The Polish Ecological Club suggests strategies for ecological restoration and nursing the forest back to health. Notably, the area does not have NGOs or forest-dwelling communities of the kind described in most of the other chapters, which is partly a legacy of the Communist past. The Polish Ecological Club explores possibilities of involving local communities in forest conservation, through ecotourism, for example.

The biologically rich Niepolomnice Forest, the focus of this case study, is located close to the industrial centres of Krakow, Tarnow and Bochnia in the southern part of the Vistula River valley, known as the Sandomierz Lowland. Here, at the confluence of the Vistula and Raba rivers, the terrain is relatively flat and the climate typical of much of Central Europe. Human settlement is limited to 20 villages in the outer part of the forest, belonging to three local communities, or *gminas*.

Historically, the Niepolomnice Forest was part of the great forest complex that linked the Carpathian Mountains to Central Poland. From the thirteenth century, the forest was subject to intense pressure and almost entirely destroyed. Fortunately, it was granted royal protection in the sixteenth century, when it was recognised as the most important source of food for the king's table in Wawel Castle in Krakow. By then, however, large fauna had been almost completely wiped out, and bears, bison and aurochs had to be imported for hunting from Bialowieza Forest. During the Partition period, the forest became the property of the Austrian Emperor and was logged to supply timber for the salt mines in Wieliczka and Bochnia. Significant damage was inflicted during the First and Second World Wars. Today there is intense air pollution from the large steel complex at Nowa Huta, located only a few kilometres to the east. There is little doubt that air pollution has caused widespread changes in the forest flora, especially noticeable in mushrooms and lichen. But the nature and extent of changes in vegetation have yet to be seriously studied. In spite of this, the Niepolomnice Forest, encompassing 110 square kilometres, remains largely intact. Most of the forest is state property, falling under the jurisdiction of a single forest office located in Niepolomnice. Although 10 per cent of the forest is privately owned, management practices in the area do not constitute a significant problem. Several hectares within Niepolomnice Forest are closed off as a bison-breeding station, managed by Ojcow National Park on the basis of a lease agreement with the State Forest Directorate.

Forest Ecology

Many of the species found in Niepolomnice Forest are listed as protected in Poland. The distinctive character of the vegetation has resulted in a designation of the area as a special geobotanical region on Polish botanical maps. Four types of forest ecosystems with a biodiversity typical of the vegetation of the whole Carpathian region occur here: multispecies lowland forests associated with *Carpinion betuli* (including stands of *Tilio-Carpinetum stachyetosum*, *Tilio-Carpinetum typicum*, and *Tilio-Carpinetum caricetosum pilosae*); fertile wetland forests associated with *Alno-Padion* (including stands of *Circaeo-Alnetum* and *Fraxino-Ulmetum*); marshy alder forests and other vegetation associated with *Alnetea glutinosae* (including stands of *Carici Elongatae-Alnetum* and *Salici pentandro-cinareae*); and mixed forests on acid soils (including stands of *PinoQuercetum*, *Molinio-Pinetum* and *Gacinio uliginosi Pinetum*).

The hornbeam forests in the northern and southern parts of the Niepolomnice are of great ecological value. These are old-growth forest stands that have been designated as nature reserves. Throughout the area many individual trees have been designated as nature monuments under the Law on Nature Conservation. These represent the remains of the forests that dominated the region during the Middle Ages. Niepolomnice Forest can thus be seen as a gene bank containing many tree species and genotypes that are found in a wide range of temperate forest ecosystems.

The flora is also quite distinct. Of particular interest are the stands of lowland birch (*Betula humilis*) in the Polanie Bloto area. There are also stands of Atlantic forest vegetation, which are found here in the eastern and southern limits of their range. Royal fern (*Osmunda regalis*) is an unusual sub-Atlantic fern also found here. Although there are many wetland species, their abundance and variety are much more limited than in the past. Signicant plants include: *Andromeda polifolia*, *Oxycoccus quadripetalus*, *Eriophorum vaginatum* and *Ledum palustre*. Mountain floral species include *Alchemilla glabra*, *Alnus incana*, *Arabis halleri* and *Veronica montana*. Approximately 30 protected species occur in significant stands, including a variety of rare mushroom species in the northern part of the forest, many of which can only be found in Poland. It is worth noting that mushrooms and lichen are extremely sensitive to air pollutants.

Field or grassland ecosystems, many with a seminatural character, form an important element of Niepolomnice. There are also several examples of marshlands that provide habitats for rare species such as *Cirsium rivulare*, *Polygonum bistorta* and *Lychnis floscuculi*. The more open-field ecosystems include buttercup and other associations (*Ranunculus acer*, *Lychnis floscuculi*), bent-grass associations (*Carici-Agrostietum caninae*), and polygonaceous associations (*Cirsio polygonetum*). All possess a rich flora, to the extent that in some northern parts of the forest there are as many as 60 different species per 100 square metres. The largest impacts on field species diversity occur at the forest edges, where the use of agro-chemicals and forestry-related activities have been most intense. Wetland vegetation is also well represented in Niepolomnice Forest, especially lowland bog plants. Despite significant changes due to human impacts, much of the marshland is protected through designation as nature reserves.

Niepolomnice Forest is one of the few remaining places in Poland where fauna which is characteristic of extensive old-growth forest can still be found. The large diversity of forest biotopes, coupled with extensive areas of forested lands and nearby field, wetland and water habitats translates also into a rich faunal biodiversity. Furthermore, several ecological corridors along the valleys of the Wisla and Raba rivers connect the area to the Carpathian mountain ecosystem and to the forests of the Sandomierz Lowland. Endangered species include the lesser spotted eagle (*Aquila pomarina*), green sandpiper (*Tringa ochropus*), short-eared owl (*Asio flammeus*), tawny owl (*Strix uralensis*) and penduline tit (*Remiz pendulinus*).

There is a high density of most mammal species occurring in Poland, including roe deer (*Capreolus capreolus*), *Muscardinus avellanarius*, and *Clethrionomys glareolus*. Old trees provide habitats for insects and reptiles. In some parts of the forest over 10 pairs of breeding birds occur per hectare – one of the highest instances in Poland.

Niepolomnice has five nature reserves extending over an area of over 100 hectares and protecting the most valued forest stands. Four of them – Lipowka, Kolo, Dlugosz Krolewski and Debina – are forested, while the fifth, Wislisko Kobyle, is an old Vistula riverbed. There are plans for additional reserves, as well as for designation of the whole Niepolomnice Forest area as a landscape park. The latter would provide less rigorous protection than a national park. The *voivode* (or state governor) is responsible for the designation of new national parks. Using extensive documentation already available, the possibility of creating a national park in the Niepolomnice Forest is currently being analysed by a working group. The Voivodship Commission for Nature Conservation (Wojewodzka Komisja Ochrony Przyrody) lent its unanimous support to the landscape park proposal in 1995.

These additional protection measures would undoubtedly help to reduce human pressure on the forest ecosystems. But they would also curtail economic and other activities in the protected area and buffer zone. These restrictions have an impact on the local people, who have already publicly expressed their concerns. Interest groups from the forestry sector have shown the greatest opposition to the landscape park proposal. As we remarked earlier, there is a conflict between forest and nature conservation authorities, which is a direct result of an inadequate division of jurisdictional responsibilities and a lack of clear goals and operational directives. This conflict is avoidable as both the forest and the nature conservation authorities are in agreement about the need to improve the effectiveness of nature conservation. Yet the designation of the area as a landscape park will do little to resolve conflicts and inconsistencies between the two authorities.

Local Forest Use and Management

A single forest office, with its headquarters in Niepolomnice, is responsible for forest management across the whole area. The forest office includes eight more local offices, with a ninth specialising in game management. Over 90 per cent of the forest is state-owned, with less than 10 per cent in private hands, of which a major proportion is owned by local governments (*gminas*). Private forest lots are scattered at about 200 sites and are largely low-quality forest used for fuelwood.

Table 15.1 Percentage forest cover by species type

Pine	65.6
Larch	1.0
Spruce	0.4
Beech	0.9
Oak	17.9
Ash	1.4
Hornbeam	0.4
Birch	1.4
Alder	10.6
Poplar	0.3
Linden	0.1

Source: Forest Office, Niepolomnice

The presence of other species, such as fir, is minimal. Average timber production is approximately 106 cubic metres per hectare, and the average age of tree stands is 42 years. Harvesting takes place on the basis of detailed management plans, clearcutting is not generally practised and, in principle, each area being logged is reforested. Annual revenues from forestry amount to 500 million old zloty (approximately US$17.000), of which a certain amount is used for afforestation. Timber from the Niepolomnice Forest is processed by local pulp mills which are located less than 100 kilometres away.

Apart from commercial logging, tree harvesting is related to maintenance of the health and quality of the forest. The basis for all forest management activities is the forest operational plan (*operat lesny*) which is approved by the General Directorate of State Forests. The long-term goal is to systematically increase the beech–oak component of the forest. Unfortunately, activities to that end are hampered by the fact that the hydrology of the area has been changing in recent years.

The professional staff in charge of the forest include forest inspectors, wardens, forest officers (25 people), an administrative staff (14 people) and forest labourers (18 people). Work related to larger endeavours is contracted out to private companies. An important part of the Forest District's responsibilities is wildlife management. In Niepolomnice Forest there are two wildlife divisions and a Regional Game Breeding Station (Osrodek Hodowli Zwierzyny Lownej). The principal game animals are deer, wild boar and stag. Frequently hunting trips are organised for foreign visitors, mainly from Switzerland and Germany. The annual revenue from hunting amounts to 700 million old zlotys (approximately US$24.000).

The Niepolomnice Forest covers an area of 287 square kilometres in three local *gminas*: Niepolomnice, Drwina and Klaj. Most of this is thinly populated, given the large percentage of forested land in the area. The forest is surrounded largely by agricultural land in individual private holdings with an average size of about 3 hectares. Locally, there are few job opportunities and the majority of local people are therefore forced to seek employment far from home, in addition to their farm work. This system of dual employment is a direct result of the centrally planned industrial development that characterised the Communist era.

Niepolomnice Municipality stands out among the *gminas* of southern Poland as one that is well-governed and has been able to create an attractive climate for foreign investment. Coca-Cola, for example, has built one of its largest bottling factories here. As a result, the *gmina* is able to generate substantial revenues that are reinvested in the further development of urban infrastructure – water, sewage, telephone systems, support to small businesses, and improvements to attract tourism. An effort is also being made to increase local employment opportunities.

As a result of recent reductions in industrial air pollution, the forest has been regenerating. The closure of the nearby Skawina Aluminium Works following public pressure by environmental groups was important in this regard. During the past three years mushrooms and cranberries have come back, which has led to a return of human foraging. But this has also placed new pressure on sensitive ecosystems and mammal populations.

One of the main challenges faced by the Niepolomnice Forest District is insect pests and fungal diseases. In 1995 old-growth forest suffered badly from leaf pests, especially pine webworm and leaf moth. Webworm threatened over 40 hectares of pine forest. Leaf moth primarily attacks oak. In 1995, it threatened over 300 hectares of oak forest. No special action was taken, however, as oak regenerates with no need for interference. Fungal outbreaks affected four of the forest's principal species and required some preventive action to hinder wider outbreaks.

Two Polish laws relate directly to forests: the Law on Forestry and the Law on Nature Protection (both 1991). In addition, some 20 other laws cover forestry-related issues – examples are the laws on land use, hunting, agriculture and environmental protection. The total forest area of Poland extends over approximately 27 per cent of the country (1996 figures), of which 18 per cent is forest land not owned by the state. The forest law contains almost no provisions for mandatory control of privately owned forest land. This means that extensive logging takes place on most private forest lands, with several examples of clearcuts, and reforestation is rarely practised. Very few *voivodes* have exercised their right under the forestry law to offer Regional Forestry Directorates the power to control forest use on private land. The Department of Nature Protection has responsibility for protected areas, including 20 national parks that come under the National Park Board (Krajowy Zarzad Parkow Narodowych). In addition, the Department shares responsibility with relevant local *voivodes* for landscape parks, of which there are over a hundred in Poland, and nature reserves.

The main environmental threat to Polish forests is industrial air pollution. As a result, there are no fully healthy large forest complexes in Poland, and a degree of forest damage can be observed almost everywhere. There may also be new kinds of diseases emerging following genetic changes in insects, a problem that has only started to receive serious attention. Two other major threats to Polish forests are from indiscriminate logging and human activities, especially uncontrolled cottage building.

It is important to note that the General Directorate of State Forests has for years been active in the conservation of large forest ecosystems, independently of the Department of Nature Conservation. Indeed, over 50 per cent of Polish forests have been designated for special protection through management practices that are meant to ensure their ecological functions. This entails the following measures:

- Introducing special provisions for ensuring forest health;
- A preference for natural forest regeneration;
- Limiting drainage;
- Promoting species most adapted to the local habitat;
- Limiting clearcutting and exploitation of resin and stumpwood.

In the past few years, foresters have been developing a new approach to forest conservation the so-called forest development complexes (*promocyjne kompleksy lesne*). There are currently five such complexes designated in Poland, including Bialowieza Forest, Knyszynski Forest and the mountain forests of Beskid Zywiecki and Silesia. Under this designation, a special type of conservation-oriented forest management is to be practised with the objective of preserving the ecosystem by limiting logging and other forms of commercial exploitation. It is important to note that the concept of the forest development complex as a conservation mechanism is not recognised in the Law on Nature Conservation. This again shows that the respective forest and nature conservation authorities are often in competition with one another.

Local Initiatives, Problems and Needs

The Niepolomnice Forest District Office has overall responsibility for Niepolomnice Forest. The state forest administration contributes only in a minor way to the budgets of local governments, mostly through a forest tax. The District is responsible for all management in state forests and assists owners of private forests in the preparation of forest management and afforestation plans. All three local municipalities are primarily agricultural in character, even though forests cover as much as 50 per cent of the area. Non-state forest lands cover only 846 hectares. These forests are predominantly very young stands of low commercial value. In the case at hand, however, the private forests are well managed by the Niepolomnice Forest District itself on the basis of special management agreements.

In this respect, it is worth noting that the regulations based on the Law on Forestry provide an opportunity for private land owners to secure tax benefits in return for reforesting land. Indeed, the current legal regulations encourage reforestation of lower-quality soils (formally, class IV, V and VI soils). This is especially significant with respect to class IV soils, which are not relieved of agricultural land tax but are relieved of all taxes if they are afforested for a period of 40 years. Moreover, local governments can receive funding for preparing forest management plans and seedlings can be provided from the Voivodship Fund for Environmental Protection.

In terms of local policy and planning, the Niepolomnice Forest is seen as a potential source of tourism-related revenues. Indeed, the development plans of the local governments associated with the Forest are focused on promoting its natural recreational values. Currently 2886 hectares, concentrated mostly in the western part of the Forest, are designated for recreation and tourism purposes by the Niepolomnice Forest District. The District has prepared car parking, camping accommodation and other tourism-related facilities as part of overall forest management.

These measures by themselves, however, are not sufficient to ensure effective protection in nature reserves, wildlife habitats or forest regeneration areas. To deliver effective nature conservation, tourism development of the Niepolomnice Forest demands much closer cooperation and co-ordination between the Forest management authorities, local offices of central government, and local government. This is largely absent at present. For example, the existing system of tourist trails is largely autonomously developed and managed by the National Tourist and Countryside Association. There is little doubt that the promotion of sustainable tourism offers a promising option to help ensure effective forest management in the area. But an integrated approach to tourism development is a precondition. The current lack of tourist trails and adequate accommodation means that weekend and day tourism dominates. At certain periods the presence of large numbers of holiday makers and mushroom pickers makes it difficult to ensure appropriate behaviour, which leads to fire and other hazards.

According to the Niepolomnice District Office, poaching of game animals and illegal tree cutting are not of significant proportions when compared with levels in other forested areas,. The relatively small percentage of privately owned land and an effective forest management system based on well-equipped forest rangers explain this successful outcome.

It is crucial that the development strategies of the respective local municipalities recognise the importance of protecting the natural values of the Forest. For this reason, there is an evident preference for promoting environmentally benign industry and ecological agriculture (especially in the Drwinia municipality). All three municipalities are committed to developing adequate sewage systems, building local sewage treatment plants and dealing effectively with the problem of waste disposal.

There is also the question of whether the environmental awareness of residents of the Forest municipalities is sufficient to warrant a development programme which accepts the conservation of the Forest's natural resources as a priority. Experience to date suggests that much can be done in this area, provided that policy formulation and implementation involves the active participation of local people. The Forest District has delegated much responsibility for environmental education to the local municipalities but it has allocated no resources for developing and promoting these activities. The District considers environmental education of low priority when compared with forest management and planning. Notwithstanding that, Forest District staff take an active part in working with schools, the provision of information to local farmers, and co-organising forest clean-ups with youth groups. There is no active environmental NGO in the area, although there is a proposal to develop a local branch of the Centre for Environmental Education in Krakow with the support of the municipality of Niepolomnice.

Securing a Future for Niepolomnice Forest

The case of Niepolomnice Forest is in many ways representative of a typical Polish forest complex. It lies close to large urban-industrial centres and is surrounded by well-managed farmland. Less than 10 per cent of the forest is in private hands, with the greater part managed by the state forest administration – in this case, the Forest District in Niepolomnice. The variety of commercial and conservation-related operations currently practised in the forest suggests that a well-coordinated management system with clear jurisdictional responsibilities and a clear vision of the forest's future is essential. Management plans are also needed, and should include systematic replanting and reforestation with the objective of changing the species mix.

Would the various management activities be sustained and improved under increased private ownership of the forest, involving several different owners? Experience since promulgation of the 1991 forestry law, which exempts private forest lands from effective regulation, suggests that from a sustainability perspective, this would not be favourable. It would be likely to result in more commercial logging, less reforestation and less planting of endemic trees.

Recognition of Niepolomnice as a large forest complex of ecological importance at the regional and national scale points to the desirability of placing it under the jurisdiction of a single authority with a clear mandate for conservation. Unfortunately, many of the proposals for restructuring and privatising forest management and forest-based industries currently under consideration appear ill-conceived from the perspective of long-term sustainability. When several institutions share jurisdiction over the same resources, conflicts are likely to arise. In a large forest complex like Niepolomnice, where surrounding villages and the proximity of large urban areas mean that human impacts cannot easily be avoided, effective coordination of conservation efforts between various central and local government authorities is essential.

The Law on Land-Use Planning outlines jurisdictional responsibilities, but it provides no procedure and little incentive for effective coordination among the various authorities. Indeed, each of the three municipalities in Niepolomnice Forest has its own vision of how to develop the area. Consequently, they are working independently in a fragmented manner with no overall coordination. It would be preferable that all major investment and infrastructural development plans be subjected to wider assessment and consultation. With the large number of land-use planning and nature conservation specialists in nearby Krakow, it should be possible to create an innovative and well-coordinated management plan for the Niepolomnice Forest. As for the proposal to create a Niepolomnice Forest Landscape Park, this might well end up as just one more bureaucracy. In Poland today there is no example of effective coordination among landscape park, municipal and forest authorities. Any changes in the existing management system, therefore, must be linked to a redefinition of jurisdictional responsibilities, especially with regard to the conservation of landscape values.

Tourism and recreation-related impacts on the Niepolomnice Forest will continue to grow. Joint management between forestry and municipal authorities is therefore necessary to minimise the impacts of hiking, biking and riding activities on the forest. Coordinated planning of marketing among the three local municipalities is also essential. It should take into account the natural and cultural heritage of the area and ensure both protection and economic viability. For conservation to be effective, local people must be encouraged to become active participants. Environmental education relating to forest conservation must be an essential component of forest management and planning. This includes adding forest conservation to school curricula.

BIBLIOGRAPHY

Agenda 21 (1992) *Forest Principles – Non-legally Binding Authorative Statement of Principles for a Global Consensus on the Management, Conservation and Sustainable Development of all Types of Forest.* United Nations Conference on Environment and Development (UNCED).

Anon. (1991) *Strategia ochrony zywych zasobow przyrody w Polsce* (Strategy for Conservation of Biological Resources in Poland), Zaklad Badan Srodowiska Rolniczego i Lesnego PAN, Oznan.

Anon. (1994) *Informacja o stanie srodowiska w wojewodztwie krakowskim w roku 1993*, Biblioteka Ochrony Srodowiska, Krakow.

Anon. (1995) *Rocznik Statystcny Wojewodztwa Kraowskiego* (Statistical Yearbook of the Krakow Voivodship), Wojewodzki Urzad Statystycny w Krakowie, Krakow.

Chiechanowicz, J. (ed.) (1994) *Ochrona srodowiska: zbior przepisow* (Environmental Protection: a Compendium of Regulations), Wydawnictwo Prawnicze LEX, Gdansk.

General Directorate of State Forests (1993–96) *Builetyn informacyjny* (Information Bulletin), Warsaw.

Glawacinski, Z and Michalik, S. (1996) *Kotlina Sanomierka* (Sandomierz Lowland), Wieddza, Warsaw.

International Union for the Conservation of Nature (IUCN) (1994) *Guidelines for Protected Area Management Categories*, IUCN, Gland.

International Union for the Conservation of Nature (IUCN) (1994) *Parks for Life: Action for Protected Areas of Europe*, IUCN, Gland (Polish summary published by Polish Ecological Club).

Jacyna, I. (1978) *Ls nie obroni sie sam* (Forests Won't Protect Themselves), Krajowa Agencja Wydawnicza, Warsaw.

Kachniarz, T. and Niewiadomski, Z. (1995) *Nowe podstawy prawne zagospodarowania przestrzennego* (New Legal Foundations of Land Use Planning), Instytut Gospodarki Przestrzennej i Komunalnej, Warsaw.

Kielczewski, B. and Wisniewski, J. (1982) *Las w srodowisku zycia clowieka* (The Forest in Human Society), Panstwowe Wydawnictwo Rolnicze i Lesne, Warsaw.

Lenkowa, A. (1981) *Zacelo sie od swietego gaju* (It Started with the Sacred Grove), Krajowa Agencja Wydawnicza, Warsaw.

Ministry of Environment, Natural Resources and Forestry. (1995) *Trawly rozwoj lasow w Polsce – stan i zamierzenie* (Sustainable Forest Development in Poland), Warsaw.

Ministry of Environment, Natural Resources and Forestry, (1996) *Polityka ekologniczna panstwa* (National Environmental Policy), Warsaw.

Radecki, W. (1991) *Samorzad terytoialny i ochrona srodowiska* (Local Government and Environmental Protection), Foundation for Local Democracy Development, Warsaw.

Sommer, J. (1993) *Prawo ochrony srodowski w Polsce – poradnik praktyczny* (Environmental Protection Law in Poland – a Practical Manual), Towarzystwo Naukowe Naukowe, Wroclaw.

Szujecki, A. (1992) *Czy lasny musza zginac?* (Must Forests Die?), Wiedza Powszechna Warsaw.

16 Revitalising Local Forest Management in the Netherlands

The Woodlot Owners' Association of Stramproy[1]

IRMA CORTEN • ETC–Netherlands
in collaboration with
NICOLE CORDEWENER • Bosgroep Limburg
and PAUL WOLVEKAMP • Both ENDS

Although almost half of Dutch forests are in private hands, the potential contribution of small woodlot owners to sustainable management and nature conservation has been underestimated considerably. According to the most recent Dutch forest survey, there are 53,000 hectares of small, privately-owned wood lots, covering 17 per cent of the country's forests (CBS 1985). This category of forest owners should therefore receive more attention from policy makers, nature conservation organisations and local authorities. For this reason we wish to take the reader to the municipality of Stramproy, where small woodlot owners and other citizens joined hands to improve forest management and enhance nature conservation. Stramproy, too, can be a source of learning and practical inspiration.

This municipality belongs to the Province of Limburg in the southern part of the Kingdom of the Netherlands. At the time that the woodlot owners began to take their initiative, income from forestry had dropped and forest holdings were being divided into increasingly smaller parcels. Many smallholders in Stramproy, as elsewhere, were becoming less involved in the management of their forest properties. Citizens and officials were thus motivated to improve the situation, and cooperation among forest owners enhanced their sense of responsibility. In order to understand the current state of local forest management in the Netherlands, we begin with a brief overview of the political-historical context and of the special role of private forest owners in the Netherlands. Then we describe the situation in Stramproy.

Forest Management in the Netherlands

With less than 10 per cent forest cover, a mere 340,000 hectares in total, the Netherlands is one of the least-forested countries in Europe. With more than 15 million people living in 43,473 square kilometres, it is also one of the most densely populated countries in the world. Forests are of minor economic importance in the Netherlands. By and large, costs of management exceed revenues. The forests are small-scale, the price of wood is low and the costs of labour relatively high. Currently, half of forest revenue comes from multi-purpose management – recreation, timber production and nature/landscape conservation. The other half is from the sale of timber alone. The state forests are managed by the semi-privatised State Forest Service (Staatsbosbeheer), although ownership by private trusts and the state has increased in the past ten years, while the share of private owners has diminished (SBH 1995). The purchase of forest land by nature conservation trusts is subsidised by the state.

Dutch cities have a major impact on forests. More than 200 million recreational visits are recorded per year (LNV 1993). Many citizens are members of, or financially support, private foundations and unions that purchase and manage forest and other natural areas for the purpose of conservation. These foundations also receive income from lotteries and donations. In other words, although they are scarce and their use to satisfy basic needs is negligible, there is a complex range of interests involved in Dutch forests.

Forest Policy

The Dutch Minister for Agriculture, Nature Conservation and Fisheries, Josiah van Aartsen presented a new policy to encourage small private landowners, like farmers and woodlot owners, to assume a more prominent role in nature conservation. This means, in a sense, that the large nature conservation trusts, which as a group are now the major beneficiaries of government subsidies for conservation, may lose their monopoly. These trusts will probably continue to manage the large, fragile and unique nature areas. But where possible and when their involvement can complement the role of the trusts and the state, private owners will become more actively involved in nature conservation. According to the Minister, this will enhance society's support for nature conservation. A serious drawback of this new policy, however, is that the government envisages substantial cuts in the budgets for the trusts – US$20 million – in order to make money available for this new target group of private land-owners (Aarden 1996).

Since 1917, Dutch forests have been protected under forest law, which introduced a few important principles, including the obligation to ensure forest generation after logging and the 'compensation principle' (added later), according to which, if forest lands are converted to other forms of land use, the same amount of additional land must be reforested (LNV 1993, Buis 1993). The annual harvest of timber from Dutch forests is 1.3

million cubic metres. This represents 10 per cent of the Netherlands' total demand, and the government aims to increase the degree of self-sufficiency in wood production to 17 per cent by the beginning of the next century. There are plans to plant an additional 75.000 hectares of forest in the next 25 years (SBH 1995 and LNV 1993).

The national government has a Forest Policy Plan, which designates most forest land as 'multi-functional forest', serving the population's recreational needs, nature/landscape conservation, wood production and environmental protection (through. CO_2 sequestration, for example). In some forest areas, nature conservation has highest priority. The government offers subsidies to promote various forest functions. To promote recreation, the government subsidises forest owners who allow the public to enter their forest (the so-called *functiebeloningsregeling*) Although this is available only for lots larger than five hectares, some small woodlot owners have applied jointly for this grant. In some cases, forest owners can also enjoy tax breaks. Those owners who possess forest with important natural values may receive an additional grant. The government categorises the land-owning nature conservation trusts as a distinct category for which a separate grant system has been created, including grants for land purchases.

A committee of forest owners and forest managers which applied international criteria for sustainable forest management to the Netherlands concluded that, whereas current legislation and policies are favourable, the vitality and biodiversity of Dutch forests is threatened by serious environmental problems such as nitrogen precipitation. The annual estimate is between 30 and 60 kilogrammes of precipitation per hectare due to air pollution from agriculture, industry and traffic. Another problem is dehydration due to the deliberate lowering of the groundwater table to cater for the agricultural sector. Approximately a third of the forest is suffering from serious drought problems (van Tol 1996). In spite of growing international attention to popular participation, the committee failed to address the role of local communities in forest management and policy.

Forest Fragmentation

Historically, there were two categories of private forest owners in the Netherlands: in the first of these was the aristocracy and, later on, rich merchants and industrialists (Wiersum 1996). Their forests either belonged to the family or were the result of investments in the reforestation of wasteland. Forests were kept both for social reasons, notably for hunting and the continuity of family traditions, and as an investment and source of income from wood production. The other category of landowners was farmers, for whom the forest was an essential element in the farming system and a source of basic needs such as fuelwood, green manure and wood for agricultural implements. The forest lost most of these functions with the modernisation of agriculture, and citizens, meanwhile, started to buy parcels of forest for recreational purposes (Wiersum 1996). For reasons explained below, few of these owners engaged significantly in management. Many of the small forest parcels became even more fragmented by multiple inheritance.

Regional Forest Groups

Many private forest owners and local authorities once received support and advice from the State Forest Service. As a result of reorganisation and ongoing privatisation, the State Forest Service has ceased to provide this sort of assistance. Though other state agencies took over the advisory role, contact with forest owners has become far less intensive. In many cases the support given is limited to specialised brochures and a few extension activities for groups. In other words, an unfulfilled demand for personal advice has developed (Corten 1991).

Partly in response to this vacuum, regional forest groups have been created in most provinces over the past 15 years. They are cooperative associations formed by forest owners, and organised at the provincial level. Membership consists of private forest owners, municipalities and, in some cases, nature conservation organisations. Their objective is to promote improved forest management and they offer their members professional support in management planning. Such cooperative arrangements bring the obvious advantage of scale and better pricing as a result of improved bargaining positions. Research shows that forest enterprises that are members of cooperative forest groups achieve better economic results (an average US$100 more per hectare/year) than those which are not associated. Active forest groups also organise courses on integrated (multi-purpose) forest management. The regional forest group in the Province of Limburg encourages small woodlot owners to establish local associations like the pilot project for small-scale woodlots in Stramproy. The forest group can best support the individual owners through these intermediaries. Until recently, the national government stimulated the creation of forest groups through subsidies (Dfl10 per 'participating hectare'). Although the government then shelved this arrangement temporarily, evaluations have been positive and financial support for non-profit activities like training and advice is likely to be continued (Bosgroep Limburg 1996).

Two Visions of Forest Management

Roughly speaking, there are two systems of forest management in the Netherlands that aim to promote nature-related values. One is integrated forest management, based on the concept of multiple forest use. The other is large-scale 'nature development'. Up to the 1970s, clearfelling was the management technique used for harvesting wood. Since then there has been a shift in thinking about forest management, reflecting the public's appreciation of the forest's ecological and social values. Heavy storms in 1972, 1973 and 1976 contributed to this change. They created open spaces in the forest where new, good-quality, forest rapidly developed. This helped to make the use of natural regeneration more common in reforestation. The work of the NGO Kritisch Bosbeheer (Critical Forest Management) and the international debates helped to pave the way for the concept of 'integrated forest management', based on mimicking natural processes and the structure of the natural forest. Selective thinning is the most important intervention that influences the forest's composition. Dead trees are left standing and wood is allowed to decay on the forest floor. The dead wood not only enhances soil formation, but is also a condition for the survival of many organisms. Some areas are also exempted from logging in order to preserve their natural values. By concentrating recreational facilities like cafes and parking lots in less fragile areas, pressure is taken off vulnerable sites. All this implies small-scale management, which allows the forest to serve different functions simultaneously (van Laar 1996, Wieman and Hekhuis 1996).

Within the Dutch context, integrated forest management is considered more favourable than traditional clearfelling from both the economic and the ecological points of view (Wieman and Hekhuis 1996). Natural regeneration is cheaper than planting, and it motivates forest managers to pay more attention to the nature function of the forest (Kuper 1994). There are of course economic limits to how much the timber harvest can be reduced, and to the number of dead trees and open spaces that can be left in the forest. Obviously, the quality of the harvested wood plays an important part. When the quality of the harvest can be guaranteed by additional management measures, however, this will compensate for reduction in yield volume, and the effect in the balance will be small.

Nature development as a forest management system evolved only recently. The primary purpose of forests managed in this way is nature conservation, with limited allowances for recreation. The system entails lending nature a hand by felling trees to open the canopy and the forest floor, for example, or by digging pools to host amphibians. To encourage the regrowth of riverine forest, arable land, previously protected by dikes, has been submerged by allowing rivers to flood their banks. Riverine forest once covered large parts of the country. Most of it was cleared by the Middle Ages.

Thousands of years ago, cows, elk and wild horses roamed the land which is now called the Netherlands. These large grazers continuously opened up spaces in the forest, thereby creating a rich variety in vegetation. In the past few years, there have been attempts to reintroduce these animals to promote a return of this type of landscape. This in turn requires large connecting nature areas. It also means that other forms of land use, such as the much appreciated and biologically rich small-scale landscapes, need to be sacrificed. Not surprisingly, this form of forest management is provoking intense debates about what should be considered 'natural' in the Netherlands in the 1990s (IKC 1996).

The Small-scale Woodlots Pilot Project in Stramproy

Forest woodlots in the municipality of Stramproy have been highly fragmented by multiple inheritance. Currently, 82 hectares of forest, consisting of the Areven (30 hectares) and the Stramproy Moor (52 hectares), are owned by about 100 people, some of whom do not know where their forest is situated or even that they possess forest. This fragmentation has hampered proper forest management over the years. As no harvesting took place, closed canopies shut out the light and hindered the undergrowth. Yet fragmentation also implied that many different types of management were practised, allowing a variety of forest types to develop.

The municipality decided to address the problem of lack of management through dialogue with a local citizens' nature conservation group and the Foundation for Preservation of Small-Scale Landscapes. The latter, Stichting Instandhouding Kleinschalige Landschapselementen (IKL), works to preserve and maintain characteristic ecological features of Limburg's landscape, such as wooded banks, pollard willows and oak scrub. Owners did not feel involved, forests did not bring owners financial benefits, and owners lacked forest management knowledge and skills. The local nature conservation group, IKL and the regional forest group initiated the small-scale woodlots pilot project in Stramproy in 1994 and approached the provincial government for financial support. This was not difficult to come by as some 25 per cent of Limburg's forests are fragmented in small-scale woodlots (Cordewener 1995). The province is currently working on a new forest policy plan, and small-scale woodlots are an important focus. The municipality contributed practical logistics and services.

In addition to the 100 private forest owners, the municipality of Stramproy and the Trust for Nature Conservation[2] own woodlots on the Stramproy Moor and the Areven. Most woodlot owners live in Stramproy village, and came by their forest through inheritance. Owners[3] that have left Stramproy maintain strong emotional ties with their forest. On the Stramproy Moor, forest is also bought by people from outside Stramproy, who visit and camp during weekends. Thus one of the preconditions for popular involvement in forest management – namely, control over the land and its resources – has been fulfilled, at least as far as the role of the forest owners is concerned.

Historical Ownership

The differing roles of the municipality and the owners can be explained in great part by historical circumstances. From the eleventh century, the municipality of Stramproy belonged to the Abbess of Thorn whose rules and decisions had the force of law. The abbess was entitled to a tenth of the harvest from agriculture and animal husbandry. In turn, she maintained the church and saw to it that services were held. The municipality's administration was entrusted to an appointed mayor. Stramproy evolved as a result of forest clearings during the early Middle Ages.[4] Making use of moors and reclaimed marshland farmers practised a mixture of land uses: forest, wild land, arable and grazing. This was known as the Esdorpen landscape. Arable land was owned by farmers, while moors, marshes, forest and grazing lands (if they had not been reclaimed by individuals) and open terrain in the centre of the village (planted with oaks) were commons. Since pre-feudal times, rights to the commons, or *gemeijnte* as they were called in the Limburg Province, belonged to the villagers living adjacent to these lands. This included rights to pasture and to wood for fuel and construction (Philips *et al.* 1965). The mayor and councils regulated the use of the commons, especially valuable species such as oak (*Quercus robur*), beech (*Fagus sylvatica*) and linden (*Tilia sp.*). Theft was punishable, and occasionally the mayor and the councils allotted wood rations to the villagers (Mandos and Kakebeeke 1971).

In 1795, following the French Revolution, the estates of the princely state of Thorn were distributed among the municipalities (Mandos and Kakebeeke 1971). Around 1835 the Areven was sold to private persons, mostly local farmers, and partly planted with deciduous forest. Other parts were used as hay fields. The Areven had previously been a marsh. To combat the wet conditions, trees had been planted on artificial earthen walls: the *rabatten*. Small canals were dug, causing the water table to drop even further. The deciduous forest is now characterised by oak and hornbeam (*Carpinus betulus*) and *elzenbroekbos*, or alder (*Alnus glutinosa*) and birch. The Areven is an old-growth habitat with natural regeneration of oak and ash (*Fraxinus excelsior*). The area has a variety of gradients and seepage water is present. This causes a rich and rare vegetation. Part of the Areven, the so-called 'poor man's land', was not sold since it was too marshy. The municipality, which remained the owner, allowed the poor to use it for cattle grazing (Natuurmonumenten 1994, Cordewener 1995).

The Stramproy Moor, where there is currently a spontaneous return of deciduous trees such as birch (*Betula spp.*), oak (*Quercus robur*) and rowan (*Sorbus aucuparia*) (Cordewener 1995), was reclaimed between 1890 and 1935 for agriculture and cattle grazing. The dryer parts were planted with pine (*Pinus sylvestris*) to grow timber for the mines and to generate employment. During the Second World War, the forest played an essential role as a provider of fuel. But over more recent decades, the economic value of the woodlots has declined sharply. The importance of the forest for the production of poles, twigs, broomsticks, construction wood and firewood has decreased. With the closure of the coal mines during the 1960s, the market demand for redwood (*Pinus sylvestris*) from the Stramproy Moor also disappeared. Thus the economic returns from the forest – another key precondition for citizen participation in forest management – deteriorated.

Nevertheless, most owners did not sell their forest. If the forest brings low returns, the costs are also low: 'the forest eats no bread', as one owner puts it. Strong emotional attachment to the forest remains, and a desire for the family to continue ownership – even if, of late, placing a caravan in one's own forest has been prohibited. Most owners give the highest priority to the natural function of the forest (Cordewener 1995). Yet most pay little attention to management. Only a few engage in timber and firewood extraction, coppicing of oak scrub, and digging of ponds for wild mammals and amphibians. The immaterial benefits are of importance. One of the owners, a retired carpenter, expressed this as follows: 'I could sell the forest and stay at home and take *valdisper* (tranquillisers), but I rather spend my time in the forest.'

A number of small-scale woodlots are owned by the Nature Conservation Trust.[5] In 1982 the Trust began buying parcels of forest in Stramproy and the Areven: it plans to practise large-scale nature development in this region and establish an undisturbed, semi-natural landscape. Wild horses and oxen will be introduced to maintain the open spaces. As Stramproy is located on the Belgian border, the trust also envisages creation of a cross-border ecological network through collaboration with Belgian municipalities and a land-owning trust. To implement this master plan, the Trust wants to buy the small-scale woodlots in Stramproy (Natuurmonumenten 1994).

The Other Citizens

The Stramproy Moor and the Areven are an important recreation area for people in the surrounding villages and for tourists. The Stramproy Moor has a camping and bungalow park. During weekends the area is popular among hikers, horseriders and cyclists because of its scenery and the fact that the area has few roads. When asked, most citizens do not want to be involved in decision making, but indicate that they are satisfied with the way the forest has been developed so far. Yet members of the local nature club accepted responsibility for the management of the 'poor men's land' in the Areven which was owned by the municipality. The flora and fauna of this marshy area are ecologically valuable. But the area was seriously threatened by 'enrichment' of the soil as the traditional practice of grass cutting had stopped. Intensive drainage of the agricultural area around the Areven was also causing the marsh to become drier. Poor, wet grasslands are very rich from a biodiversity point of view. The nature club saved the 'poor men's land' by removing turf sods. At times, however, there were not enough volunteers to guarantee adequate management. It was therefore decided that the municipality should hand over the area to the Trust. Members of the same nature club also lobbied the municipality to enhance collaboration with and among the forest owners to ensure more coherent management. This encouraged the municipality to support the initiative that led to the Stramproy small-scale woodlots pilot project.

The Pilot Project Commences

Establishing an association of woodlot owners required identification of all owners, a labour-intensive exercise due to extensive fragmentation of land. The complicated ownership pattern – often single forest areas had multiple heirs – partially explains why an association did not evolve spontaneously among forest owners. The initiators of the project – the municipality, the IKL, the nature club and the regional forest group – played an indispensable catalyst role.

The project's coordinator, Nicole Cordewener, who is affiliated to the Regional Forest Group Limburg, commenced a dialogue with groups such as Nature Conservation Trust and the local nature and environment groups. These consultations also helped to create a positive environment in which it became possible to discuss issues that the different groups and persons involved felt to be important. The municipality provided overall support – for example, it assisted the association in identifying forest owners in its records. A process of discussions, exchanges of information and awareness raising then commenced, involving notably the woodlot owners themselves. Plans developed towards the establishment of the woodlot owners' association. At an early stage, 43 owners agreed to collaborate with the association. Six owners decided to sell their forest to the Trust. In general, one can say that the pilot project encouraged the forest owners to reflect on the plans they may have for their forest.

After consultations with the owners, the project coordinator drafted a management plan for the Areven and the Stramproy Moor which was eventually approved by the owners. At the same time, discussions began with the Trust regarding its nature development management plan. This entailed three basic strategies for management of its lands: human-made landscape; semi-natural landscape; more-or-less natural landscape. The Trust's management of the Areven has so far focused on the preservation of a small-scale human-made landscape – grassland, hayfields and indigenous types of forest. Mowed grass is removed to allow poorer soils and richer vegetation to develop. Pools have been dug on the grasslands, and were soon populated by amphibians. In order to check drainage and abate the problem of dehydration in this area, ditches in the forest and in the grasslands are being dammed (Natuurmonumenten, 1994). On the Stramproy Moor, a return of endemic forest is being promoted. The woodlots are currently dominated by pine. Through heavy thinning of pine stands, the regrowth of deciduous forest is stimulated. The Trust aims to achieve a semi-natural landscape consisting of large tracts of forest relieved by open terrain (Natuurmonumenten 1994).

Ironically, the Trust's participation in the pilot project – and hence its contribution towards strengthening the association – could undermine the Trust's own interest. The more successful the association, the smaller chances are that owners will sell their forests. Already many owners are unwilling to sell their land. It is thus important for all parties to involve the owners in efforts towards improved management.

The Pilot Project's Forest Management Plan

Forest management is based on the concept of integrated forest management. The plan aims to sustain the forest so that it contributes to nature conservation, wood production, recreation, scenic beauty and a healthy environment. These objectives are to be fulfilled simultaneously, with the condition that management should be self-financing. Sustainability implies that the costs are kept to a minimum, making forest management less dependent on economic circumstances. The plan offers a coherent framework, with enough flexibility to allow the individual owners to achieve their own specific objectives. With the plan as a guide, it is the forest owners who will in the end make decisions about their forest land (Cordewener, 1995).

Management of the Areven forest aims to allow old-growth forest to recover through natural regeneration, in order to convert forest on the Stramproy Moor from pine (*Pinus sylvestris*) to mixed deciduous forest with oak (*Quercus robur*) and birch (*Betula sp.*). This will enhance ecological values and recreational possibilities in the region. The management plan also refers to national wood production and the country's responsibility to the international community. In this respect, trees with a large diameter will be grown, since the timber yield of these trees will not interfere with ecological objectives. There is also a relatively stable market for large diameter hardwoods (Cordewener 1995).

Organization

A committee has been formed consisting of nine owners and a representative from the municipality. As it does not consider itself mature enough to register legally as an independent foundation or cooperative, the Regional Forest Group Limburg continues to hosts its secretariat, and prepares the annual workplans and newsletters. The Regional Forest Group Limburg finances these activities from its own resources. The group hopes individual forest owners will eventually run an independent association.

The association organises meetings and guided excursions with the assistance of local nature groups. Through the media, citizens are invited to take part in these events, which increases general knowledge about the forest.

Management activities have begun with the felling of exotic species such as Canadian poplar (*Populus canadensis*) and *Prunus serotina*, which has become a pest in the Netherlands. The municipal Public Works Department offered its services free of charge.

The pilot project encouraged the owners and other inhabitants of Stramproy to give more thought to the future of their forest, with non-owners actively encouraged to participate. The fact that the municipality – not to mention the local school teacher and the baker – owned woodlots added to the project's success. All forest owners, including the municipality, have appreciated the management support offered by the Regional Forest Group Limburg, especially since the State Forest Service stopped offering this sort of assistance. Greater attention is now paid to silvicultural aspects such as the selective felling of high-quality timber, which previously was only converted into firewood. The association could also enable forest owners and other citizens to join hands against external threats, such as dehydration of the soil.

Conclusions and Recommendations

The Netherlands is party to the Biodiversity Convention and the Forest Principles of Agenda 21, which were drafted during the United Nations Conference on Environment and Development in Brazil in 1992. These two agreements give priority to local people's participation and the support of society at large for the protection of the natural environment. The approach adopted towards forest management in Stramproy offers a concrete example of a feasible strategy to achieve popular participation in the use and protection of forests. Such local cooperative efforts deserve to be supported.

Stramproy demonstrates two approaches with regard to the role of citizens in sustainable forest and nature conservation. The association focuses notably on local people, whereas the Nature Conservation Trust aims at the wider public. The association strives for integrated forest management, whereas the Trust aims at large-scale nature development. Such differences need not hinder collaboration between the Trust and the association. Problems like drought and over-manuring, for example, can be combated jointly.

Looking at Dutch forest management in a historical perspective, the circumstances for participation have changed dramatically. The economic returns and services available to forest owners have dropped. Fragmentation of woodlots makes it increasingly difficult to ensure adequate management. The developments in Stramproy teach us that such obstacles, which create a growing sense of impotence, can be removed through collaboration expressed in membership of the association and the Regional Forest Group.

A number of factors have been decisive in ensuring the association's success. It is of the first importance that the owners themselves determine whether or not certain measures are introduced in their forest. In other words, they maintain control over their forest. The association adds economics of scale and, consequently, brings the owners financial gain. The cost of harvesting and maintenance is reduced; owners are eligible for government subsidies; and they are likely to receive better prices for their timber. Moreover, the owners have the possibility of learning more about the forest ecosystem and various management options. Advice and support from the regional forest group and the information provided by local nature groups add to the owners' skills and knowledge.

Because the association has good relations with other organisations and involves as many citizens as possible, the project has a positive impact on society. Above all, it is well-rooted in the local community. The financial, political and moral backing of the initiative by the province and the municipality is crucial, as is the Limburg Regional Forest Group's catalyst role. And the forest group's uninterrupted assistance and support for the initiative remain, even at this stage, indispensable. It is strongly recommended that the government continues to support the work of the regional forest groups.

The experience from Stramproy thus suggests some essential preconditions that may determine the success of citizens' efforts at sustainable forest management, and also provides a confirmation of forest management experiences in the tropics (Laban 1994).

- Citizens must have the feeling that they can influence outcomes (claim-making power).
- People should be able to fish, to hunt, to extract timber or to harvest

other products from the forest – but also to enjoy nature, to learn about the ecosystem, and to experience the practical fruition of their own vision of the forest and nature management (material and immaterial gain).

- People should have the ability to manage the forest, to interact with policy makers, to analyse present conditions and to plan for the future (knowledge and skills).
- People should have security about their formal positions in the process, and the acknowledgement thereof by others (rights).

Other important preconditions are adequate financial means and support from local authorities, as well as through policies and regulations. Ideally, it is the task of the national government to create the conditions that can enable citizens to take responsibility for using and managing the natural environment in a sustainable way (Laban 1994). Experiences from Stramproy, like those from the tropics, show us that genuine participation can contribute to sustainable forest management. Participation helps to improve efficiency, to avoid conflicts and to remove social obstacles. It is hoped that the experiences of successful initiatives such as the pilot project in Stramproy will be shared with forest owners and others throughout the Netherlands.

Looking Forward by Looking Back: An Afterword on the History of Forest Management in the Netherlands

A historical review of forest management in the Netherlands is perhaps the best way to understand the ecology of Dutch landscapes and the involvement of rural people in the management of their natural environment. It is difficult to say which types of forest originally belong here. The climate changed drastically after the last Ice Age. Since then, people have changed the landscape and soil structure by removing peat, clearing forests, developing and reclaiming waste lands, and poldering lakes. Even before the time of Emperor Charlemagne in the seventh century, most old-growth forests had disappeared (Buis 1993). All remaining forests in the Netherlands are young.

Between the ninth and the thirteenth centuries, wood extraction for use in iron smelting threatened the forest. This forced the local population to preserve nearby forests for the use of their own community. Local institutions called *marken* evolved to govern these forests. In the *marken* not controlled by a feudal lord, there was a voting system related to ownership of a farm. User rights over the forest and 'wild lands' were essential for the farm and the satisfaction of the people's basic needs. The members of the *marken* harvested wood (primarily for fuel), grazed their animals, collected leaves and cut sods of turf for manuring their agricultural fields. They also processed oak bark for tanning and hunted small game (Buis 1993).

The local *marken* institution provided the basis for small-scale forest and landscape management. Management measures were introduced, such as 'leaving the land in peace' (fallow), planting trees and bushes, and coppicing. The problem of overgrazing was a constant concern. Although rules for the fallowing of grazing land were agreed upon, many inhabitants violated them. Tree planting, an obligation of all members of the *marken*, was often unsuccessful due to the shortage of seeds and seedlings and because of unchecked grazing (Buis 1993).

Between the tenth and sixteenth centuries, major parts of the country's landscape were dominated by a highly integrated farming system, traces of which are still discernible today. This is the case in Stramproy. The cattle stable was the central component of the mediaeval farming system. Turf sods and leaves from the forest were brought to the stable as bedding, to which the cattle added their dung. Year after year, a rich mixture of manure was added to the agricultural fields where cereals were grown. The ratio between farmland, moorland (heather) and grazing land, which was determined by the fertility of the soil, was more or less fixed. To change this ratio could seriously endanger the system and lead to erosion (Buis 1993). Thus the 'esdorpen landscape' evolved where agricultural fields (*essen*) were surrounded by wild lands and forest.

As a result of overgrazing and uncontrolled collection of leaves, forests degraded and turned into heather, while the ongoing cutting of heather for turf sods created wind-blown sands and large, eroded, unproductive waste lands (Buis 1993). Landlords, who viewed the forest as a money box, were also responsible for some destruction (Buis 1993). During the 1750s total forest cover in the Netherlands was only 4 per cent of the land area. The use of the wild lands for manure production continued to restrain the possibilities for afforestation until the arrival of artificial fertiliser in the 1850s.

Cities influenced the country's forests as early as the sixteenth century, when the Netherlands constituted one of the most urbanised and densely populated countries in Europe. City dwellers had lost daily contact with the forest, and no longer perceived it as a source of basic needs. Instead, forests were increasingly appreciated for their recreational function. For example, when Prince William of Orange planned to sell the Forest of The Hague ('Het Haagse Bos') in order to pay his war debts, the well-to-do citizenry of The Hague furnished a large sum of money to preserve the forest for recreational purposes. The Forest of The Hague remains to this day (Buis 1993). During the sixteenth and seventeenth centuries many of the newly rich patricians – such as successful merchants – desired a country seat, and estate forests often surrounded castles and country houses. This was also the time of the discovery of 'new' continents: exotic plants were imported to decorate the gardens and estate parks. The oldest estate forests were usually located on the better soils.

From the mid-eighteenth century, the same class that had planted estate forests argued that wild lands and supposedly unproductive moorlands should be cleared for the purpose of growing cereals. Yet the soils of the wild land and moors were not fertile enough to grow cereals, and there was not enough animal manure to fertilise them. So short-rotation forests were used instead. It was believed that after some 25 years these forest plantations would have produced enough leafy manure to make it possible to fell the trees and convert the land to high-quality agricultural land. This did not work out well, however, and the forest remained. Some attempts were even made to reafforest the dunelands, but these also failed (Buis 1993).

The communal lands, or *marken*, disappeared at the beginning of the nineteenth century as a direct result of liberal legislation. The state divided the *marken* and, where possible, cleared the wild lands to increase welfare. The enactment of the new *marken* law in 1886 was the *marken's coup de grâce*. It ruled that a single user of the *marken* could demand its division, and led to the fragmentation of the remaining *marken*. In a short time, most land was up for sale. There was a general consensus that the *marken* institutions did not have the capacity to clear and develop the wild lands and moors.

The trend towards privatisation was boosted by the import in 1868 of fertiliser, which made it possible to reclaim wild lands. The agricultural crisis – the corn deficit and high fuel prices – all contributed to large-scale reclamations. Railway and tram connections established in 1886 facilitated the transport of fertilisers, plant material and large volumes of timber for the mines. Furthermore, rich citizens could more easily leave their cities and travel by train to their estates. This stimulated the development of new estates, and more forests were planted as a form of investment. The state felt it had to play a role in reclaiming the land and it established the State Forest Service. (Staatsbosbeheer) (Buis 1993).

From the end of the First World War and into the 1930s, large employment schemes were undertaken to abate unemployment, including the afforestation of moorlands. Local authorities and private foundations and associations could apply to the state for advances free of interest (Buis 1993).

The first land-owning nature conservation organisation, the Trust for Nature Conservation, was inaugurated in 1905, its establishment prompted by the City of Amsterdam's plans to convert the Naarder Lake into a waste dump. This motivated a number of people to mobilise public opinion and financial means to save important nature areas, like Naarder Lake and wild lands and moors which would otherwise have been reclaimed (Natuurmonumenten 1956). Smaller nature conservation organisations were established at the provincial level (Buis 1993).

The youngest forests are in the polders. Recently farmers have, with the help of state subsidies, been encouraged to plant 'farmers' forests'. This has the twofold objective of extending forest cover and reducing agricultural overproduction. These farmers-cum-forest managers are, in general, more active entrepreneurs than most traditional forest owners. They capture various state subsidies such as premiums for CO_2 sequestration and drought abatement Most traditional forest owners, including the state, take a rather passive approach – even though many amongst them face serious economic problems. Forests and nature are still poorly represented in the country's state budget (Buis, 1993, Wiersum, 1996).

NOTES

1 The authors wish to thank the following people for their valuble contributions and advice: Harrie Weersink, Bosgroep Limburg; Toon van den Eijnde, Natuurmonumenten; Jacques van Dael; forest owners and citizens from Stramproy; the municipality of Stramproy; Huub Ruijgrok and Jan Willem Meurkens, Small World Media; Arend-Jan van Bodegom and Kees van Dijk, IKC-NBLF; Freerk Wiersum, LUW; Tjeu Moors, Gemeentewerken Stramproy; Peter Laban, ILEIA; Henk Lette, ETC.

2 The Vereniging Natuurmonumenten is a private land-owning trust which manages around 70,000 hectares in various parts of the country. Legally it is registered as an association. With approximately 830,000 members, it is the largest nature conservation organization in the Netherlands. The Trust aims to promote the preservation and restoration of nature and landscapes.

3 Although most forest woodlots are registered under the name of men (or husbands), women take an active interest in forest management in Stramproy, through excursions, meetings, and to a lesser degree, the actual management. Only one woman, however, is member of the association's committee. But then the coordinator of the pilot project, who is employed by the Regional Forest Group Limburg, is also a woman.

4 'Roy' in the name Stramproy means open spot in the forest.

5 The members of the Trust can influence the general policies through representatives elected on a district level who have a vote on the Trust's council and appoint the board. Important decisions of the council have included the decision to limit hunting and to restrict the public's access to the nature areas. Members have little say, however, over day-to-day management. This is entrusted to specialists employed by the Trust. In other words, participation is synonymous with direct representation, and is restricted mainly to the level of the overall policy. The Trust informs its members about the nature areas through a magazine and a detailed guide. Most areas are open to the public.

BIBLIOGRAPHY

Aarden, Marieke (1996) 'Ook boeren moeten natuur gaan beheren. Van Aartsen roept verzet op bij professionele organisaties', *De Volkskrant*, 15 October.

Aartsen J. J. (1996) Speech, 21 October.

Bosgroep Limburg (1996) Newsletter, November.

Buis, J. (1993) *Holland Houtland, een geschiedenis van het Nederlandse bos*, Amsterdam.

Centraal Bureau voor de Statistiek (CBS) (1985) De Nederlandse bosstatistiek – deel 1 (1980-1983).

Cordewener, N. (1995) *Documentatie map, behorende bij de eindrapportage van het proefproject "samenwerking bij het beheer van kleinschalig bosbezit" in gemeente Stramproy*, Bosgroep, Limburg. Nuenen.

Corten, I. (1991) *Het voorlichtingskundig begrip `kennis systeem 'toegepast op de sector bosbouw, globale omschrijving van het kennis systeem; elementen, hun relaties en taakafstemming*, Department of Educational Sciences, Wageningen Agricultural University.

Corten, I. (1992) 'Het kennis en informatie systeem van de Nederlandse bosbouwsector', *Nederlands bosbouwtijdschrift*. Vol. 64, No. 4 (July), pp. 135–42.

Corten, I., P. Laban and L van Veldhuizen (1996) *Ervaringen met lokale participatie, een verkennend onderzoek op het werkveld bos en bosbeheer*, ETC-Nederland and IKC-Natuurbeheer.

IKC (1996) 'Themanummer begrazing: grazen als biologisch fenomeen en als beheersmaatregel', *Bosbouwvoorlichting*, No. 35, September, pp. 120.

Jehae, M. (1996) 'Van Aartsen houdt het op dynamiek', *Natuur en Milieu*, October, pp. 810.

Kuper, J. H. (1994) 'Sustainable development of Scots Pine forests', PhD thesis, Wageningen Agricultural University.

Laar, J. van (1996) 'Van dennenakker tot bosecosysteem, een historische schets van anderhalve eeuw bosbeheer in Nederland', *Inzicht*, Vol. 1, No. 8 (January), pp. 710.

Laban, P. (1994) *Accountability, an Indispensable Condition for Sustainable Natural Resource Management*, ETC-consultants Leusden .

LNV (Ministerie van Landbouw, Natuurbeheer en Visserij) (1993) *Bosbeleidsplan, regeringsbeslissing*, LNV, Den Haag.

Mandos H. en A.D. Kakebeeke (1971) *De acht zaligheden, oude kern van de Kempen, Bijdragen tot de studie van het Brabantse heem*, Part 12.

Natuurmonumenten (1994) *Beheerplan 1994*, Weert/Stramproy.

Pater, de, C. (1996) 'Participatie bij duurzaam bosbeheer: van Pakistan tot de Oostermoer', *Nederlands Bosbouwtijdschrift*, December.

Philips, J. F. R., Jansen, J.C.G.M. and Claessens, Th. J. A. H. (1965) *Geschiedenis van de Landbouw in Limburg 1750–1914*, Assen.

SBH (1995) *Nederlands bos in beeld*, Stichting Probos, Zeist.

Tol van G. (1996) 'Internationale ontwikkelingen en de betekenis voor Nederland', *Bosbouwvoorlichting*, 1996, No. 3.

Vereniging tot Behoud van Natuurmonumenten in Nederland (1956) *Vijftig jaar natuurbescherming in Nederland.*

Wieman, E. A. P. and Hekhuis, H. J. (1996) 'Geïntegreerd bosbeheer: beter voor de natuur, ook beter voor de portemonnee?', *Inzicht*, Vol. 1, No. 8 (January).

Wiersum, K. F. (1996) 'Particuliere bosbouw in Nederland: Een nieuwe toekomst?', *Nederlands Bosbouw Tijdschrift*, January.

Wijnia, B. A. and Houwaard, D. H. (1995) *Beslissingen in het bosbeheer, een verkennende studie naar bedrijfsstijlen in de bosbouw*, Working Paper IKC-Natuurbeheer No. W-91, Wageningen.

17 Tigers, Mushrooms and Bonanzas in the Russian Far East

The Udege's Campaign for Economic Survival and Conservation

JENNE DE BEER and
ANDREY ZAKHARENKOV
Far Eastern Association for the use of Non-Timber Forest Products in
Khabarovsk

Forest and Forest Policy in the Russian Far East

With a total forested area of about 7.64 million square kilometres, the Russian Federation accounts for over 22 per cent of all the forests on the world's surface. The Russian Far East covers 6.63 million square kilometres of land surface, some 45 per cent of which is covered by forests. In this remote corner of the world, many of the last pristine, wild salmon runs are found (Pacific Environment Resource Centre 1996). The region is sparsely populated, with only about ten million people living in this huge area, most of them in the southern part. Less than 200,000 people are indigenous to this part of Russia.[1]

The forest industry[2] provides almost 10 per cent of the Russian Far East's total industrial production and forms the social fabric of many communities, with entire villages built up around logging, hunting and other forest uses. Since 1989, timber production has been declining, which has increased economic and social hardships for many of these communities (World Bank 1996). Reasons for the decline include a decrease in federal subsidies, shrinking domestic demand and rising transport and fuel costs. In Soviet times, timber extraction was organised to meet production quotas and little attention was given to improving efficiency and logging techniques. Today, tremendous waste still plagues the timber industry. Between 40 and 60 per cent of all timber cut is lost in the production process, which is increasingly oriented towards the raw log export market. Foreign investment in the timber sector by companies from Japan, South Korea and elsewhere appears not to be directed at processing. Developing locally based processing and manufacturing value-added products would provide more income per tree, reduce waste, and benefit local communities by providing jobs. But this is not being done. Rather, by focusing on the short-term gain from raw log exports, timber companies are speeding up logging and, faced with the growing scarcity of accessible stands, are looking to develop road-less wilderness. A large part of the forests in the Russian Far East is inaccessible due to the rugged landscape and poor infrastructure. The forest industry has already heavily overlogged many of the more accessible areas, however, particularly around railroads and near population centres.

The southern part of the Russian Far East has the most productive and biologically diverse forests. The Sikhote-Alin mountain range and coastal forests along the Sea of Japan, with some of the richest forests in the Russian Far East and situated close to ports, are currently among the most urgently threatened areas. With rising transport costs and strong demand from countries such as Japan, South Korea and China, the pressure on these relatively accessible forests is growing.

The Sikhote-Alin mountain range

The biologically richest forests of the Russian Far East are to be found in the Ussuri *taiga* along the Sikhote-Alin mountain range. The mountains stretch from southwest to northeast in Primorsky Krai, well into Khabarovsk Krai, parallel to the coastline. The area still has a number of relatively intact watersheds along the Bikin, Samarga, Khor, Chuken and Anui rivers.

The mixed coniferous-deciduous forests of the mountain range are renowned for their rich biodiversity. The forest contains an estimated 3,000 higher plant species and is recognised by the International Union for the Conservation of Nature (IUCN) as a Centre of Plant Diversity. Furthermore, the forests provide the habitat for a unique fauna which includes rare species such as the Amur tiger, the Chinese merganser, Blakiston's fish owl and several endangered species of crane. The Amur tiger ('Siberian tiger') is under threat from poaching and habitat destruction. Only an estimated 250–400 tigers are left throughout the southern part of the Russian Far East (Gordon 1995).

The Udege

The Udege are one of the indigenous peoples who have inhabited Russian Asia since ancient times. Today, the Udege number about 1,600 people. They live in the northern part of Primorsky Krai and in the southern part of Khabarovsk Krai, and their livelihood depends almost entirely on the forest. The Udege economy revolves around hunting, fishing and the collecting of forest products. Only in this century did they take up agriculture, and even today this remains a very marginal activity. In the past fish, in particular *taimen* and salmon, formed a very important part of the Udege diet.

Fish stocks have decreased rapidly over the last decades, hauled down by pollution and logging down river, and by over-fishing in the Ussuri River, of which the Bikin is a tributary. Meanwhile, over the years, the Udege diet has shifted more and more from traditional food towards 'modern', processed food, which reportedly has affected general health conditions negatively.

The animals chiefly hunted for the pot are deer and wild boar,[3] the main game for commercial hunting are fur-bearing animals: sable, mink and several type of species of squirrels. Dried deer antlers are sold on the black market for large sums of money and there is a strong Chinese and Korean demand for the gall bladders of bear and for *kabarga* (*Moschus moschiferus*) musk, both widely used in the Chinese pharmacopia. With the current buying power of the Korean and the Chinese, black market prices are soaring. Hunting pressure on *kabarga* and bears is far beyond sustainability and both species are being driven to extinction locally.[4] The Bikin is one of the very last refuges of the Amur tiger, much respected by the Udege who strictly avoid killing this animal. Unfortunately the poachers who fly in by helicopter do not share these scruples.[5]

Before 1958 the Udege lived in small settlements scattered over the Bikin Valley. In 1958 most of the Bikin Udege were resettled in the village of Krasny Yar. Today, the population of the village consists of 700 individuals, including a few dozen people belonging to related indigenous groups (such as the Nanai and Oroche) and some Russians. In the traditional culture of the Udege respect for nature was an essential value (Arsenjew 1924). While the Soviet administration brought them various improvements, such as better health care and education, it also damaged the indigenous culture (Vakhtin 1992). As a consequence, today much of the Udege culture, including traditional knowledge relating to the management of forest resources, has been lost. Nevertheless, a core of the culture and way of life is still intact, as exemplified by the important role forests continue to play in daily life, and the deep respect for the tiger. The Udege are currently trying to revitalise such core elements in their culture.

Parts of the traditionally indigenous areas currently have the status of Territories of Traditional Nature (TTNs). This arrangement provides indigenous people with a voice in the resource management decisions affecting the lands upon which they are dependent. Until recently, some of the most important old-growth forests lacked any form of protective legal arrangement. In 1998, however, the breakthrough occurred when two new protected areas were established: one area of about 700,000 hectares on the Bikin River and a second one of about 300,000 hectares on the Chuken River. Both protected areas were strongly lobbied for by Udege people. As it is now, the central and most important part of the Sikhote-Alin mountain range appears to be protected.

Several logging companies, however, are trying to expand the range of their operations and now pose an immediate threat to these forests. Among the logging companies active in the region, two stand out. The Korean mega-corporation Hyundai has been logging the area since 1992 and has established a joint venture with the Russian logging company Primor-LesProm. Having already contributed significantly to the destruction of nature in the Far East,[6] Hyundai now wants access to the western side of the Sikhote-Alin mountains, targeting more than 300,000 hectares of forest in the Bikin River watershed. So far, the Udege of the Bikin have strongly and stoutly opposed this scheme. But Hyundai is not giving up and continues to lobby the regional government to open up the Upper Bikin to logging. A relative newcomer is Malaysia's Rimbunan Hijau (RH). The company has recently acquired a 305,000-hectare long-term lease in the Sukpai River watershed. The allowable cut it has secured is 550,000 cubic meters per year. RH estimates that timber volumes from the given territory are likely to be lower than those originally announced by the administration, and has asked for nearby forest to be reserved for logging in order to provide the promised cut to the company. The impact of RH's logging operations are likely to extend even further than the leased area. The company plans to build a road from the town of Sukpai to the port of Nelma to export the timber. The road would bisect the Samarga River watershed, which contains 800,000 hectares of roadless ancient forests and has been recognised by local scientists as a 'biodiversity hotspot' (Gordon 1998).

While opposed to the opening up of their forests to large-scale logging operations, the Udege and other local people wish to develop their economic activities based on the forest. These activities focus on the rich non-timber forest product (NTFP) resources of the area. Major products are mushrooms and ferns, pine nuts, birch juice, Siberian ginseng (*Eleutherococcus senticosus*) and berries, among which the excellent *limonik* (*Schizandra chinensis*). Various leaves are collected as ingredients for health teas. Another, related, product is honey. A very fine quality of honey is produced in the area. The bees feed on rich nectar sources from ginseng and other medicinal plants in a pollution free region.

Before the collapse of the Soviet Union, several *GosPromKhozes* (state forest product companies) and *KoopzveroPromkhozes* (cooperative forest product companies) organised the trade in NTFPs from the region. After 1989 a new situation appeared. A tax regime now virtually discourages legal productive activities. In the aggregate, taxes may be well over one hundred per cent. In the past, forest products were delivered according to a pre-set plan at fixed prices. In addition, at the time, the quantity of products was more important than the quality. Now, primary producers face the challenge of exploring markets for their products – finding new customers on the home market and linking up with trade networks for organising the export of products. What made the situation worse was that locally produced items were driven out of the home market and replaced by imported goods. Many plants for processing berries and medicinal plants had to close down. Prices of raw materials came under pressure and, because of inflation and a fluctuating market, it became very difficult to arrange long-term delivery contracts with customers. The existing trading companies turned out to be ill-equipped to deal with this new situation. As a result, the incomes of those dependent on the enterprise fell dramatically.

Current Efforts to Improve the Economic Situation

In the early 1990s it became clear, that action had to be taken. Several Udege village organisations on the Khor River in Khabarovsk Krai and on the Bikin River in Primorsky Krai became involved in efforts to develop forest-related economic activities on their own. The programme they started is focusing on the marketing of NTFPs and on the development of ecotourism and is now operational at community and regional levels. At the community level, activities are directed at building an organisational infrastructure that can handle the collecting, packaging and transportation of NTFPs from the collecting sites. The same organisational structure is to deal with tourism activities. Recently such activities were also taken up by the Udege living along the Samarga River.

At the regional level, the Russian Far East Association for the Use of Non-Timber Forest Products has been established. The association, based in the town of Khabarovsk, has as its primary task to ensure sustainable development in remote forest villages and to support the traditional economies of the region's indigenous peoples through the promotion of NTFP activities. It is the aim of the association to link different actors in the region involved in the NTFP trade. These range from village organisations to traders, processing plants and scientific institutes. The association further advises and supports village organisations in establishing primary processing units, adequate storage and packaging facilities and other infrastructural needs. Jointly, the organisations can build the strength to promote their products, both on the home market and abroad. Already their common trade mark, a young Amur tiger, is becoming quite well known in the Russian Federation. Finally, the association is planning to assist people in other NTFP-producing regions, such as the Republic of Altai, to establish support organisations.

The Potential for NTFP Development in the Bikin Valley

The main NTFP product groups in the Bikin Valley and adjacent areas fall into five groups:
- Medicinal plants;
- Fruits and nuts;
- Mushrooms and ferns;
- Animal products; and
- Essential oils and resins.

Medicinal plants

There is still a big market for medicinal plants in Russia itself. At the same time, a renewed interest in traditional medicine in Asia and the expanding market for health products and homeopathic and phyto-therapeutic products in the countries of the West provide potential new outlets for a number of forest plants. In the Bikin Valley large stocks of Siberian ginseng (*Eleutherococcus senticosus*) occur. Its leaves, roots and berries can be used in tonics and other health products. During summer the plant's roots and leaves are collected in the forest close to the village.

Fruit and nuts

Pine nuts (*Pinus koraiensis*) are harvested close to the river banks in the middle Bikin pine nut reserve. There exists a steady domestic demand for pine nuts. The Bikin Valley is also a rich source for the lovers of forest berries. *Brusnika* berries (*Vaccinium vitis-ideae*), for example, are an excellent fruit for juices and marmalades. They grow abundantly in the upper-Bikin. The harvest (which is very labour-intensive) takes place at the end of October and in early November, when the hunters are preparing themselves for the hunting season. The leaves are also collected by being cut off at ground level, which stimulates the growth of new sprouts and leads to larger and better berries afterwards. Good stocks of *limonik* (*Schizandra chinensis*), a twining small shrub, grow at easily accessible places in the Bikin. Limonik is sold as whole berries or as juice and seeds. There is a steady demand on the domestic Russian market. *Limonik* juices or syrups are highly valued as a tonic. *Shipovnik* bushes (*Rosa canina* = dog rose) grow abundantly in the area and are very rich in vitamins. The plant is common in many other areas in Russia, but demand for the berries on the home market is strong, as it is a very popular fruit. Although the leaves of this plant are currently not harvested in the Bikin, they are widely used in Russia in tea mixes. In general there is great potential for the processing of berries and leaves in the production of herbal teas, in particular health teas.

Mushrooms and ferns

Two fine edible ferns occur in the Bikin Valley: *orlyak* (*Pteridium aquilinium*) and osmund (*Osmunda japonica*). The latter fern is not only in demand as a delicacy; in addition, it is said to have the property of removing radioactive contamination from the human body. For this reason *osmund* ferns are a popular side dish in the area around Chernobyl these days. A great variety of edible mushrooms grow in the Bikin area, particularly in the upper-Bikin. Mushrooms are collected for local consumption in the village and for marketing elsewhere in dried or marinated form. In a good harvest year, there is a mushroom bonanza, when many people from small towns come to the upper-Bikin to collect mushrooms (mainly for their own consumption).

BIBLIOGRAPHY

Arsenjew, Wladimir, K. (1924) In der Wildnis Ostsiberiens, 2 vols, August Scherl Verlag, Berlin.

Beer, J. H. de (1993) The Udege Forest Enterprise: Problems and Perspectives, report written for the Udege Association in Krasny Yar, Russian Far East; RIC/PERC, Amsterdam.

Beer, J. H. de (1995) The Bikin Revisited, Old Problems, New Perspectives, Sacred Earth Network/ Rainforest Information Centre, Amsterdam.

Dudley, N. (1992) Forests in Trouble: a Review of the Status of Temperate Forests World-wide, World Wildlife Fund (WWF), Gland.

Gordon, David (1995) 'Roar of the taiga', Common Future, Autumn.

Gordon, David (1998) 'And so it begins ... Russian forest tendered to Malaysian TNC', Taiga-News, No. 23, (January), p. 1.

IWGIA (1990) Indigenous Peoples of the Soviet North, IWGIA Document No. 67, Copenhagen.

Izmodenov, A. G. (1993) 'Inventory of Productive Flora of the Russian Far East', Khabarovsk, unpublished.

Newell, J. and Wilson, E. (1995) The Russian Far East: Forests, Biodiversity Hotspots, and Industrial Developments, Friends of the Earth-Japan, Tokyo.

Pacific Environment Resource Centre (1996) 'Salmon Restoration and Protection in the Russian Far East', programme update.

Vakhtin, Nikolai (1992) Native Peoples of the Russian Far North, Minority Rights Group International, London.

World Bank (1996) Russian Federation Forest Policy Review, Promoting Sustainable Sector Development During Transition, 2 vols, Washington.

NOTES

1 This figure rises to 600.000 if the Yakut people are included

2 An overview of the state of forest and forest policies in the Russian Far East is given in Newell et al. 1995. The following discussion builds on this publication

3 Manchurian chestnut (Juglans mandshurica) occurs abundantly in the upper-Bikin. The nuts are an important source of food for the wild boar in the area. They are also harvested for fodder for domestic pigs in the village.

4 Both the brown bear and Asian black bear are hunted for their gall bladders. Particularly the black bear is threatened

5 The poachers are after the bones, testicles and other parts of the tiger, which are very much in demand in China and in Korea as an ingredient in medicines. The local authorities lack enough personnel and sufficient equipment to guard this large area and fight the poachers

6 The destructiveness of the joint venture's logging operations in the Sikhote-Alin mountains has been documented on video by Dr Konstantin Mikhailov.

Appendices

Appendix 1 (Chapter 4)

Major Flora and Fauna of Economic Importance Found in the Research Areas of the Cordillera Mountains, Philippines

Local name	Scientific name	Remarks
A. GAME ANIMALS		
1. *Sabag*	*Gallus gallus*	
2. *Banyas*	*Varana salvator*	
3. *Motit*	*Viverra sp.*	
4. *Ugsa*	*Cervus mariannus*	None in Demang, Bugang
5. *Laman/buka*	*Sus barbarus*	None in Demang, Bugang
6. *Colasisi*	*Loriculus philippinensis*	
7. *Kuyat*	*Various types incl. migratory*	
8. *Bakes*	*Macaca fasciulares*	
9. *Buwet*	*Phloemys cumingi*	
10. *Paniki*	*Ptyropus vampyrus*	None in Demang, Bugang
11. *Liplipot*	*Macroglossus lagochilus*	
12. Squirrel		
13. Wild cat	*Felix minuta*	
14. Wild rat	*Rattus sp.*	
15. *Koop*	*Bubo virginianus*	
16. *Martines*	*Acridotheres cristatelus*	
17. Snake	*Ptuophis sp.*	
18. *Alimaong*		
19. *Bugatan*		
20. *Pokaw*		

Local name	Scientific name	Remarks
B. MEDICINAL PLANTS/FRUITS		
1. Dael	Pittosporum resiniferum*	
2. Subosob	Blumea balsamifera	
3. Dangla	Vitex negundo	
4. Gayyab	Psidium guajava	
5. Lomboy	Syzygium cumini	None in Demang, Bugang
6. Dalipawen	Alstonia scholaris	
7. Amti	Solanum nigrum	
8. Pinit	Rubra fraxinofolia	
9. Bugnay	Antidesma bunius	None in Demang, Bugang
10. Katinge	Premna odorata	
11. Degway	Saurani	

Appendix 2 (Chapter 5)

Alternatives to Rainforest Logging in a Chachi Community, Ecuador

Table A2.1 Animals currently hunted in the *chacras* or cropping areas

Common name	Scientific name
Raton liso	Proecimys semiespinosus
Guanchaca	Caluromys derbianus
Guanta	Agouti paca
Tatabra	Tayassu pecari
Conejo	Sylvilagus brasilliensis
Sajino	Tayassu tajacu
Perdiz	Tinamus major
Pavas de monte	Penelope purpurescens
Loras	Amazona farinosa
Paloma montanera	Geotrygon veraguensis
Pichilingo	Pteroglossus erythropygius
Paleton	Ramphastos brevis

Table A2.2 Products for self-consumption

Common name	Scientific name
Arroz	Oryza sativa
Maiz	Zea mays
Cana de azucar	Saccharum officinarum
Frijol	Phaseolus vulgaris

Papa china	Xanthosoma sagittifolium
Rampira	Carludovica palmata
Achiote	Bixa orellana
Pina	Ananas comosus
Naranja	Citrus sp.
Guaba	Inga sp.
Papaya	Carica papaya
Chontaduro	Bactris gasipaes
Caimito	Pouteria caimito
Fruta de pan	Artocarpus altilis
Aguacate	Persea americana
Maracuya	Passiflora edulis
Granadilla	Passiflora ligularis
Badea	Passiflora quadrangularis
Mandarina	Citrus reticulata
Sapote	Pouteria sapota
Guineo	Musa sapientum
Tamarindo	Tamarindus indica
Palmito	Euterpe chaunostachys
Zancona	Socratea sp.
Coco	Cocos nucifera
Tagua	Phytelephas aequatorialis
Chirimoya	Annona cherimola

Table A2.3 Animal food

Common name	Scientific name
Zorro	Didelphis marsupialis
Armadillo	Dasypus novemcinctus
Mico	Cebus capucinus
Oso hormiguero	Tomandua tetradactila
Venado	Mazama americana
Tigre	Panthera onca
Guatuza	Dasyprocta punctata
Gavilan	Herpetohteres cachinnans
Guacharaca	Ortalis erythroptera
Carpintero grande	Dryocopus lineatus
Tigrillo negro	Felis yagouaroundi
Ocelote	Felis pardalis
Mongon	Alouatta villosa
Guacamayo	Ara severa

Table A2.4. Construction timber

Common name	Scientific name
Cana guadua	Guadua angustifolia
Tagua	Phytelephas aequatorialis
Palmito	Prestoea trichoclada
Coco	Cocos nucifera
Rampira	Carludovica palmata
Piquigua	Heteropteris integerima
Cedro	Cedrella odorata
Guayacan pechiche	Minquartia guianensis
Amarillo	Persea ringens
Laurel	Cordia alliodora
Laguno	Vochysia macrophylla
Tangare	Carapa guianensis
Pambil	Socratea sp.
Sande	Brosimum utile
Chanul	Humiriastrum procerum
Cuangare blanco	Diacranthera sp.
Chalviande	Virola sp.
Chalviande peludo	Virola duckei
Anime	Dacryodes occidentalis
Peine de mono	Apeiba membranacea
Guagaripo	Nectandra sp.
Mascarey	Hyeronima chocoensis
Sangre de gallina	Vismia obtusa
Gualte	Wettinia utile

Table A2.5 Wild fruits

Common name	Scientific name
Uva de monte	Pourouma cecropiifolia
Madrono silvestre	Rheedia acuminata
Caimitillo	Chrysophyllum sp.
Naranjilla silvestre	Solanum sessiliflorum
Palmiche	Euterpe chaunostachys
Palmbil	Iriartea deltoidea
Chapil o mil pesos	Janusia caudata
Palma real	Attalea colenda
Guaba	Inga sp.
Guayacan	Tabebuia guayacan
Mani de Arbol	Arachis hypogea

Table A2.6 Fibres

Common name	Scientific name
Piquigua	*Heteropsis integerima*
Yare	*Carludovica funifera*
Hoja blanca	*Maranta sp.*
Chocolatillo	*Ischnosiphon arouma*
Damagua	*Poulsenia armata*
Cocedera	*Cecropia sp.*
Matamba	*Desmoncus sp.*
Guinul	
Chapilillo	*Jessenia bataua*
Mocora	*Oenocarpus maporea*

Appendix 3a (Chapter 6)

Trees Utilised in the Yvytyrusu Range of Paraguay: Characteristics and Potential Uses

1. *Aguai* — *Chrysophyllum gonocarpum (Mart. et Eichler) Engl.*
   Fruits; wood.
2. *Amba'y* — *Cecropia pachystachya Trécul*
   Pioneer plant; wood for paper pulp; accoustic insulation; medicinal
3. *Amba'y guasu* — *Didymopanax morototoni (Aubl.) Decne. et Planch.*
   Wood for laminated boards; packing; accoustic insulation.
4. *Apeyva* — *Heliocarpus popayanensis H.B.K.*
   Light wood for laminated boards.
5. *Cancharana* — *Cabralea canjerana (Vell.) Mart.*
   Fine wood for furniture; laminated boards; medicinal; melliferous.
6. *Cedro* — *Cedrella fissilis Vell.*
   Fine wood for furniture; carpentry; laminated boards; medicinal; melliferous.
7. *Chipa rupa* — *Alchornea triplinervia (Sprengel) Muell.Arg.*
   Pioneer plant; wood for paper pulp; laminated boards; packing.
8. *Guajayvi* — *Patagonula americana L.*
   Wood for rural construction; laminated boards; carpentry; scaffolding; melliferous.
9. *Guapo'y* — *Ficus enormis (Mart. ex Miq.) Miq.*
   Wood for packing; scaffolding; medicinal; melliferous.
10. *Guatambu* — *Balfourodondendron riedelianum (Engl.) Engl.*
    Fine wood for furniture; carpentry; laminated boards; floors; melliferous.
11. *Guavira pytã* — *Campomanesia xanthocarpa Berg.*
    Fruits; wood for rural construction; tool handles; medicinal; melliferous.
12. *Incienso* — *Myrocarpus frondosus Allemao*
    Fine wood for furniture; carpentry; construction; floors; medicinal; melliferous.

13. *Inga guasu* *Inga uruguensis Hook. et Arn.*
Pioneer plant; fruits; wood for packing; scaffolding; construction; melliferous.

14. *Ka'a oveti* *Luehea divaricata Mart.*
Pioneer plant; wood for furniture; frames; floors; tool handles; shoe-making; melliferous.

15. *Ka'a vusu* *Lonchocarpus muehlbergianus Hassler*
Wood for floors; rural construction; melliferous.

16. *Kamba aka* *Guazuma ulmifolia Lam.*
Pioneer plant; fine wood for furniture; carpentry, laminated boards; shoe heels; forage; melliferous.

17. *Karóva* *Jacaranda micrantha Cham.*
Wood for carpentry; laminated boards; packing; medicinal.

18. *Kupa'y* *Copaifera langsdorffii Desf.*
Wood for scaffolding; laminated boards; medicinal.

19. *Kurupa'y kuru* *Anadenanthera colubrina (Vell.) Brenan*
Pioneer plant; wood for construction; floors; posts; tannine; melliferous.

20. *Kurupa'y ra* *Parapiptadenia rigida (Benth.) Brenan*
Pioneer plant; wood for construction; floors; tannine; melliferous.

21. *Laurel guaika* *Ocotea puberula (Nees et Mart.) Nees*
Wood for laminted boards; construction; packing; melliferous.

22. *Laurel* *Nectandra angustifolia (Schrader) Nees et Mart. ex Nees*
Wood for construction; laminated boards; melliferous.

23. *Laurel moroti* *Ocotea diospyrifolia (Meisn.) Mez*
Wood for rural construction; fine furniture; carpentry; laminated boards; paper pulp; packing; melliferous.

24. *Laurel sa'yju* *Nectandra lanceolata Nees et Mart. ex Nees*
Wood for carpentry, rural construction; laminated boards; melliferous.

25. *Loro blanco* *Bastardiopsis densiflora (Hook. et Arn.) Hassler*
Pioneer plant; wood for carpentry; rural construction; laminated boards.

26. *Manduvi ra* *Pithecellobium saman (Jacq.) Benth.*
Wood for rural construction; floors; forage (leaves and fruits); melliferous.

27. *Mbavy* *Banara arguta Briq.*
Wood for rural construction; fine furniture; melliferous.

28. *Mbokaja* *Acrocomia totai Mart.*
Fruits; wood for rural construction; forage (fruits and leaves); medicinal; melliferous.

29. *Palmito** *Euterpe edulis Mart.*
Wood for rural construction; edible heart.

30. *Parapara'i guasu* *Pentapanax warmingianus (Marchal) Harms*
Pioneer plant; wood for paper pulp; laminated boards.

31. *Peterevy* *Cordia trichotoma (Vell.) Arrab. ex Steud*
Fine wood for furniture; carpentry; laminated boards; boats; rural construction; melliferous.

32. *Peterevy Moroti* *Cordia glabrata (Mart.) A. DC.*
Wood for laminated boards; carpentry; packing.

33. *Pindo* *Syagrus romanzoffiana (Cham.) Glassman*
Fruits; wood for rural construction; forage (fruits and leaves); medicinal; melliferous.

34. *Samu'*   *Chorisia speciosa St.Hil.*
Very light wood, for laminated boards (as part of the internal sheet); accoustic insulation; packing; fibre.

35. *Tajy hu*   *Tabebuia heptaphylla (Vell.) Toledo*
Wood for carpentry; construction; floors.

36. *Takuara*   *Bambusa guadua Humb. et Bonpl.*
Wood for rural construction; fences.

37. *Tatajyva*   *Chlorophora tinctoria (L.) Gaud.*
Fruits; wood for fine furniture; carpentry; posts; construction; floors; melliferous.

38. *Tembetary*   *Fagara naranjillo (Griseb.) Engl.*
     sa'yju   Wood for carpentry; laminated boards; medicinal.

39. *Timbo*   *Enterolobium contortisiliquum (Vell.) Morong*
Wood for carpentry; scaffolding; boats; packing; forage; melliferous.

40. *Trebol**   *Amburana cearensis (Allemao) A.C.Smith*
Wood for fine furniture; carpentry; laminated boards.

41. *Uruku rä*   *Croton urucurana Baill.*
Wood for chips; melliferous.

42. *Urunde'y mi*   *Astronium urundeuva (Allemao) Engl.*
Wood for rural construction; posts.

43. *Urunde'y para*   *Astronium fraxinifolium Schott*
Wood for construction; floors.

44. *Yerba mate*   *Ilex paraguariensis St.Hil.*
Infusions; medicinal; widely utilized

45. *Yrupe rupa*   *Guarea kunthiana A. Juss.*
Fine wood for furniture; carpentry; melliferous.

46. *Ysapy'y guasu*   *Machaerium paraguariensis Hassler*
Wood for carpentry; laminated boards.

47. *Yvaporoity*   *Myrciaria rivularis Cambess.*
Fruits; wood for rural construction; medicinal; melliferous.

48. *Yvapuruvu**   *Schizolobium parahybae (Vellozo) Blake*
Pioneer plant; wood for packing; scaffolding; laminated boards; melliferous.

49. *Yvyra ita*   *Lonchocarpus leucanthus Burkart*
Wood for rural construction; laminated boards; melliferous.

50. *Yvyra ju*   *Albizia hassleri (Chodat) Burkart*
Not apt for dense forests; but very good for planting in orchards and agriculture plots, since it provides light shade and fertilized the soil; firewood.

51. *Yvyra pepe*   *Holocalix balansae Micheli*
Wood for carpentry; rural construction; posts; tool handles; floors.

52. *Yvyra pere*   *Apuleia leiocarpa (Vogel) Macbr.*
Wood for carpentry; floors; laminated boards; floors; scaffolding.

53. *Yvyra pytä*   *Peltophorum dubium (Sprengel) Taubert*
Pioneer plant; wood for construction; floors; frames; melliferous.

54. *Yvyra ro*   *Pterogyne nitens Tul.*
Pioneer plant; wood for floors; carpentry; construction; vehicle frames.

Species for protection of streambanks

1.	Takuara	Bambusa guadua
2.	Loro blanco	Bastardiopsis densiflora
3.	Kupa'y	Copaifera langsdorfii
4.	Peterevy morotï	Cordia glabrata
5.	Uruku rä	Croton urucurana
6.	Yrupe rupa	Guarea kunthiana
7.	Apeyva	Heliocarpus popayanensis
8.	Inga guasu	Inga uruguensis
9.	Ka'a ovetï	Luehea divaricata
10.	Yvaporoity	Myrciaria rivularis
11.	Guajayvi	Patagonula americana
12.	Guavira pytä	Campomanesia xanthocarpa
13.	Kamba akä	Guazuma ulmifolia

Melliferous species

1.	Cancharana	Guarea canjerana
2.	Cedro	Cedrella fissilis
3.	Mbokaja	Acrocomia totai
4.	Guapo'y	Ficus enormis
5.	Guatambu	Balfourodondendron riedelianum
6.	Guavira pyta	Campomanesia xanthocarpa
7.	Guajayvi	Patagonula americana
8.	Incienso	Myrocarpus frondosus
9.	Karóva	Jacaranda micrantha
10.	Ka'a vusu	Lonchocarpus muehlbergianus
11.	Kamba akä	Guazuma ulmifolia
12.	Kurupa'y kuru	Anadenanthera colubrina
13.	Kurupa'y rä	Parapiptadenia rigida
14.	Tajy hu	Tabebuia heptaphylla
15.	Laurel	Nectandra angustifolia
16.	Laurel sa'yju	Nectandra lanceolata
17.	Laurel morotï	Ocotea diospyrifolia
18.	Laurel guaika	Ocotea puberula
19.	Manduvi rä	Pithecellobium saman
20.	Mbavy	Banara arguta
21.	Peterevy	Cordia trichotoma
22.	Pindo	Syagrus romanzoffiana
23.	Tapia guasuy	Alchornea triplinervia
24.	Tatajyva	Chlorophora tinctoria
25.	Timbo	Enterolobium contortisiliquum
26.	Yrupe rupa	Guarea kunthiana
27.	Yvaporoity	Myrciaria rivularis
28.	Yvyra ita	Lonchocarpus leucanthus
29.	Yvyra ju	Albizia hassleri
30.	Yvyra pytä	Peltophorum dubium

---

* Introduced species

Sources: Brack, W. and Weik, J. H. (1993) *El bosque nativo del Paraguay, Riqueza subestimada.* DGP/MAGGTZ, Serie No. 15, Asunción.